George P. Mitchell

NUMBER TWENTY-SIX
Kenneth E. Montague Series in Oil and Business History
Joseph A. Pratt, General Editor

George P. Mitchell

Fracking, Sustainability, and an Unorthodox Quest to Save the Planet

Loren C. Steffy

TEXAS A&M UNIVERSITY PRESS COLLEGE STATION

First edition

This paper meets the requirements
of ANSI/NISO Z39.48–1992
(Permanence of Paper).
Binding materials have been chosen for durability.
Manufactured in the United States of America

Library of Congress Control Number: 2019908922
ISBN 13: 978-1-62349-803-0 (cloth: alk. paper)
978-1-62349-804-7 (Ebook)

To Mom and Dad,

for nurturing curiosity, cultivating ambition,

and believing in dreams.

Whatever you can do or dream you can, begin it.
Boldness has genius, power, and magic in it.

—Johann Wolfgang von Goethe, quotation found among George Mitchell's private papers

A Note about "Fracking"

Many geologists and petroleum engineers object to the spelling of "fracking." Because the term is short for "fracturing," they argue it should be spelled without the "k." Indeed, early scientific papers written about the technique often refer to the need to "frac" a well, or the process itself as "fracing" or even "fraccing." Phonetically, though, "fracing" would be pronounced "frace-ing." As the technique entered the public lexicon, the "k" was added, in keeping with the tenets of English, and that is a practice I have continued throughout this book.

Unfortunately for the industry, "frack" is also a euphemism for an expletive on the 1970s science fiction show *Battlestar Galactica*, and when that show was revived in 2004, the term was resurrected. The show's popularity coincided with the widespread use of fracking. For many environmental groups, the irony was too delicious to ignore, and "fracking" became a derisive term applied to almost any form of drilling for oil or natural gas.

When the myths and hyperbole are stripped away, fracking has benefits as well as drawbacks, but for George Mitchell, it was part of a lifelong effort to make the world a better place.

Contents

Preface

The C. W. Slay #1 juts upward through sparse, open prairie, surrounded by a chain-link fence. The wellhead itself is only about six inches in diameter, capped by a steel valve painted a drab gray green and faded from years of scorching North Texas sunshine. Far to the southeast, barely visible on the horizon, are the skyscrapers of downtown Fort Worth.

Mitchell Energy and Development Corp. drilled the natural gas well in 1981, piercing the Barnett Shale formation about 7,500 feet below the scrubby surface. The company operated hundreds of other wells in the area, but this one was different. This one changed the world, although it would be two decades before anyone—including company namesake George P. Mitchell and his geologists and engineers—realized it.

Today, to workers in the neighboring gas fields, the C. W. Slay #1 is something akin to Graceland for Elvis fans. The Railroad Commission of Texas, which regulates the state's oil and gas industry, requires companies to post a sign declaring the name of each well, the amount of acreage in the lease, the identifying number that the commission assigns to it, and an emergency phone contact. As the fracking boom raged across North Texas from 2006 to 2014, thieves who were obviously well versed in industry lore repeatedly stole the metal marker on the C. W. Slay #1.

To those unfamiliar with the energy business, the well is unremarkable. It's hard to pick out from among dozens of similar wellheads nearby. But with the C. W. Slay #1, Mitchell Energy, a little-known, midsized natural gas producer, took the unconventional first steps that would shake up global energy markets as dramatically as the Middle Eastern oil embargoes that dominated the 1970s. The C. W. Slay #1 and the subsequent wells drilled into the Barnett formation laid the foundation for the "shale revolution," proving that natural gas could be extracted from the dense, black rock thousands of feet underground.

Fracking, the process for unlocking those gas reserves in commercial quantities, came later. It would be almost a quarter century, in the late 1990s, before George Mitchell and his team perfected the process that transformed the energy landscape.

The path to the discovery that unlocked vast, untapped reserves of natural gas—and later, oil—across the country was far from a straight line, and it was cluttered with the contradictions that defined George Mitchell's life. He was an oilman who protected habitats for endangered whooping cranes, a wildcatter who befriended renowned British physicist Stephen Hawking, the father of ten children who fretted about overpopulation, the billionaire who haggled with a Walmart clerk over the cost of fishing gear, an entrepreneur who was slow to pay bills but who shared his wealth generously with employees and charities, a demanding boss who was known for shouting matches with his second-in-command yet was approachable and even shy with others. Among his papers were two cartoons that he found amusing and that his children said defined his attitude. The first shows a long-legged waterfowl swallowing a frog. Even though the frog's head is inside the bird's beak, its hand is around the bird's throat. The caption reads "Don't ever give up." The second shows two buzzards sitting on a tree branch. One says to the other: "Patience, my ass! I'm gonna kill something!" George Mitchell's career is a testament to not giving up, and while he could demonstrate remarkable patience, he believed in creating his own opportunities.

He was intensely competitive, not in his desire to beat someone else but in his drive to demand more of himself and excel at the interests he pursued. An avid fisherman, he always strived to catch the biggest speckled trout or redfish he could; a lifelong tennis player, he prided himself on a well-placed shot; and each run he skied in Aspen, he tried to shave a few seconds off his best time or perfect the smooth angulation taught to him by Austrian ski instructors. Even the one time he played croquet, in his brother's backyard, he wanted to make the best shot. "You have always looked at life as a challenge to be enjoyed and lived to the hilt," his brother-in-law, Alando Jones Ballantyne, wrote to him in 1984.[1]

"Dad was that classic entrepreneur—highly leveraged, but highly optimistic," his oldest son, Scott, said. George Mitchell followed his interests, and his desk was littered with as many as a dozen projects that he worked on simultaneously. "He would expect other people to understand you don't work on one project and stay with it until it's finished; you work on this project and this project and this project all at the same time and give it enough attention until it's finished," his longtime assistant Linda Bomke said. "And it was never really finished, because for everything he took off the to-do list, he added two more things." When the world's largest telescope project needed funding, he paid for it; when his alma mater needed a new physics building, he underwrote it; when his hometown languished in economic decay, he led the effort to revitalize it; when researchers wanted to build the world's largest particle accelerator, he helped them launch it. When he felt that promising biotech research needed more funding to unlock breakthroughs, he set up a venture capital company to fund it. He believed that business leaders had a responsibility beyond their own balance sheets. They should provide leadership on pressing, long-term global issues as well.

He could be highly critical of his colleagues in the energy industry, which he once described as "brilliant, tough, secretive and hated."[2] While he made his fortune from fossil fuels, it bothered him for decades that the world remained so dependent on them. "We should be working on alternatives," he told a reporter in 1991. "But unless you get a reasonable price, alternative sources won't work."[3] At the time, wind and solar power cost almost double what oil did, which he argued would lead to economic disaster. The natural gas produced with fracking would do as much as any technological advancement to break that economic barrier. Cheap natural gas, unleashed by Mitchell's pursuit of fracking, provided reliable and affordable support for renewables. When the sun didn't shine or the wind didn't blow, natural gas–powered generators could kick in to offset the lost power faster and more affordably than plants powered by coal or nuclear energy.

Any one of Mitchell's many projects would have kept most businessmen occupied full time, but he spread his attention from the depths of the earth to the farthest reaches of space and managed to keep it all straight in his head. His mind was a miraculous filing cabinet from which he could recall the tiniest detail of a given project. If he saw a worker in the hall, he would ask about the latest data on a specific well. He was notorious for calling employees on Sunday night and firing off orders in a crisp, staccato cadence. A month might go by, but when he saw them again, he would expect a detailed update.

The profits of Mitchell's oil company funded these myriad interests. But if you'd told him in 1981 that one of his far-flung pursuits would change the world, he would have thought you were talking about The Woodlands, his master-planned community north of Houston. The 28,000-acre spread, which he began assembling in the 1960s, served as a workshop for his unwavering obsession: solving the social ills facing America's cities. Many oil companies invested in real estate, but none developed a single property with such attention to detail and environmental foresight. For George Mitchell, it was the "development" side of his Mitchell Energy and Development Corp.—rather than the energy side—that sparked his truest passion. But in the 1980s, the "energy" part of the company needed a solution to counteract declining reserves in its biggest gas field. Mitchell found himself racing against time to find new sources of natural gas before the field sputtered out. Convinced of a radical prospect—that new energy could be coaxed from the shale rock his drillers had ignored for years in favor of shallower, more pliable formations—he employed a team of geologists and petroleum engineers to figure out how to unlock the would-be treasure. The prospect of fracking, though still untested and unproven, emerged as the best bet to supply it. Over the years, he sank about a quarter of a billion dollars that Mitchell Energy couldn't spare into what was little more than a science experiment.

Mitchell could indulge in such persistence, largesse, and some might say recklessness only because he had voting control over his company's stock. For investors, buying a share of Mitchell Energy wasn't so much a bet on an oil and gas company as a financial testament of faith in Mitchell's uncanny knack for finding gas and his soft-spoken persuasiveness in lining up lucrative deals. Fracking would test that faith—among shareholders, employees, and the firm's own board of directors. And it would force Mitchell to split his company and ultimately sell the part he prized the most.

Today, "fracking" detractors have made the term shorthand for irresponsible drilling. The hard consonants of the word give it the ring of a curse, and the label has been used indiscriminately to describe things that have nothing to do with the process. There are no "fracking rigs" or "fracked oil" or "fracking pipelines." Wells that are fracked are drilled with much the same equipment as any other; the hydrocarbons they produce are no different from the oil and gas that flow from the thousands of wells drilled for decades without fracking. Fracking comes after the drilling, when water, sand, and chemicals are injected into the well bore under high pressure to make tiny cracks in the shale. The sand props those fractures open to allow the natural gas or oil to escape.

Supporters see fracking as proof of American ingenuity, of the triumph of free enterprise over the relentless fear of energy scarcity that defined US policy for four decades beginning in the 1970s. In the wake of fracking's hard-won financial success and the recent revitalization of US energy production, George Mitchell is often portrayed by industry supporters as the lone hero, the misunderstood visionary who risked his own fortune against all odds. Mitchell was indeed a visionary, but the money he gambled wasn't all his—his love of debt and Mitchell Energy's two-tiered stock structure gave him access to large pools of outside capital. And while the success of fracking is often attributed to Mitchell's perseverance, it is also a tribute to his abject stubbornness, his voracious inquisitiveness, and healthy doses of luck. Fracking's day, when it finally arrived, came in the twilight of Mitchell's long career. He did not develop it on his own, of course. In fact, no one even recalls him visiting the site of the C. W. Slay #1 or any other frack job his company did. But he created a culture that encouraged employees to try things, to fail, and to try again. That laboratory environment, a manifestation of Mitchell's personality and his belief in the lengthy time horizons that characterized his biggest projects, would cast doubt on the company's future. But his approach would also allow a team of dedicated employees to pursue Mitchell's vision unimpeded—and to reach his goal.

"We weren't a big company; we weren't Shell," former Mitchell Energy spokesperson Tony Lentini said. "But we did something important. You couldn't have done that at a big company. We didn't have a lot of bureaucracy, and you could talk to the top guy."

◈

The first time I talked to the "top guy," I didn't know what to expect. I had been the business columnist for the *Houston Chronicle* for a few years when I got the call. "George Mitchell wants to talk to you. He wasn't very happy with what you wrote." It was Dancie Ware, Mitchell's longtime publicist. By then, in 2006, George had sold his energy company and maintained a private office in the Chase Bank Tower, the tallest building in downtown Houston, just across the street from the *Chronicle*. I don't recall what I wrote in the column, only that it touched on this new drilling process known as fracking.

You can't publish opinion pieces about business in a business-driven city like Houston if you aren't prepared for chief executives to yell at you. I didn't know much about George Mitchell, other than he had started his own oil company and developed The Woodlands, where my wife and I lived. I knew longtime residents had a fierce devotion to "Mr. Mitchell." After my wife had minor surgery and was about to leave the hospital, I asked her doctor which driveway I should pull my car into to pick her up. He flipped over a piece of paper, drew a schematic of the hospital, and launched into a lengthy discussion of the "genius of Mr. Mitchell's design." Although the hospital was close to the interstate, its parking lots and driveways faced inward, away from the feeder roads. "Mr. Mitchell" had planned it this way to improve traffic flow and access—just one of the countless details that reflected his foresight and commitment—and the doctor's enthusiasm reflected the reverence that many residents still hold for the town's founder. Mitchell created The Woodlands from scratch in 1974, and what it lacks in history it makes up for in pride in every detail that even today makes the community unique.

For my audience with "Mr. Mitchell"—most former employees still call him that—I walked across Texas Avenue and rode the elevator to the thirty-sixth floor, bracing for the worst. I'd been criticized, yelled at, and threatened by my share of executives. It wasn't fun, but it was part of the job. I planned to listen patiently, take what I could from the conversation, and make a polite exit the first chance I could.

It soon became clear this wasn't going to be a chewing out. As I entered George's office, he came out to meet me, steering his electric scooter through the doorway to the small reception area and smiling as he shook my hand. He apologized that he couldn't stand to greet me. His knees were shot. An avid tennis enthusiast since he was twelve years old, he had played almost every day of his life, winning amateur tournaments well into his seventies. He had founded a prominent Houston racquet club, financed the tennis center at his alma mater, Texas A&M University, and reigned among the best players in the state. But all those years of playing had wrecked his knees.

Despite the physical hardship, George still came into the office daily. He was eighty-seven years old by then. He had lost most of his hair at an early age—he was bald before he was forty—but his head was still ringed with some wisps of gray. Unlike many CEOs, George was at ease talking with reporters. He spoke softly and seemed perfectly comfortable with gaps of silence in the conversation. (While he liked talking to reporters, he hated having his picture taken. For years, Mitchell Energy used the same photo of him—in his trademark coat and tie, seated behind his desk—because he refused to have his likeness reshot.)[4] Ware hadn't coached him on how to interact with me. He didn't need "media training" or talking points, and he didn't seem concerned about staying "on message." Instead, we had as casual a conversation as one could have with George. He seemed interested to hear we had gone to the same school, Texas A&M, but then there was a pause. Still smiling, he turned, pulled out a book, and handed it to me. It had a gray cover with the old Mitchell Energy logo—a representation of the sun rising over mountains. He wanted me to have it, he said, because it would help me understand fracking and why it was important. *The Barnett Shale Play: Phoenix of the Fort Worth Basin; A History* was written by Dan Steward, who coordinated the effort that unlocked the secret to successfully fracking shale rock. It's written by a geologist for geologists, but it documents the history of how fracking unfolded. George had signed the copy, as had Steward. I still have it, and it proved invaluable in writing this book. I left the office with no better idea than when I arrived about what in my column had bothered him. Seven years later, Ware didn't recall, either. Perhaps he was correcting a mistake I had made, or perhaps he simply wanted me to be better informed. Despite years in the bare-knuckle business of independent oil exploration, George didn't enjoy conflict. He cared less about pointing out someone's error than he did about increasing that person's understanding.

Business biographies typically focus on strategy. We want to know how a famous CEO achieved success in hopes that we can adopt those same methods in our careers or daily lives. George Mitchell's strategy, however, is unlikely to be replicated. He took a long view of his projects and was willing to lose money—often large amounts—over decades as he pursued his ideas. Fracking is one example. The Woodlands is another. Elements of its design, such as the use of green space, flood control, and pedestrian walkways, have been copied by other master-planned communities nationwide. Many of those developments lack George's attention to detail or the long-term view that made The Woodlands a success. Developers are less willing to give up valuable land for green spaces or walking trails, and they are more likely to level all the trees before they build. Most public company CEOs would have been fired by their boards for spending so much on projects that had so little promise of a near-term return. In both energy and development, George carved out a unique means of indulging his dreams at shareholder expense, an unorthodox arrangement that ultimately allowed his ideas to touch anyone who lives in a planned community or who has bought a tank of gas.

President Barack Obama, in his 2012 State of the Union address, took credit for the government's role in developing fracking. The oil industry pushed back, saying that free-market innovation made the technique possible. In truth, neither the government nor business could have—or would have—developed fracking on its own. Few companies would have invested so much for so long with so little to show for it. The government expanded the knowledge of shale formations and the techniques for extracting natural gas from them, but that insight alone wasn't enough to make fracking profitable.

It took George Mitchell, his foresight, and his confidence that his engineers would find a way to extract gas from shale. It also took the unique circumstances in which he found himself in the early 1980s—sitting atop a prolific gas field that was playing out and being under pressure to meet a contract that had been his company's cash cow for decades. A timely and unexpected boost from rising natural gas prices helped tip the scales in fracking's favor.

Both fracking and sustainable development were born from George Mitchell's vision, steadfast determination in the face of unyielding adversity, and an unwavering faith in his own direction and the abilities of people closest to him. George was asked once if it took "inner strength" to persevere when everyone said he was wrong. "Well," he replied, "it's that, or you're so naïve you don't realize it won't work."[5]

The inherent mismatch of his life's work—as a champion of both responsible development and nonrenewable energy—didn't seem to concern him. To him, the two were related. He understood that a continuous and affordable energy supply was a prerequisite for sustainability. Other vital supplies like food and water can't get to people without energy. Ultimately, his strategy was less about business success and ruminating on his legacy than about the pure satisfaction of celebrating ideas—even if they were as seemingly contradictory as blasting fossil fuels out of shale buried deep in the earth and building a sustainable city above.

George P. Mitchell

CHAPTER 1

A Farewell to Trees

A bald man in a brown suit, his shoulders slumped, stepped into the humid air of the Texas morning. He hobbled slightly as he walked down the short set of steps, his knees complaining from a lifetime of pounding on the tennis court. He trudged toward his waiting Cadillac Seville. It was mid-June 1997. The stifling blanket of heat and humidity that marked the Gulf Coast summer had not yet settled in, but the trees that surrounded the driveway and enveloped the entire resort and conference center behind him managed to create a stillness that echoed the emptiness he felt inside. He loved the trees. Of everything he had accomplished as a developer of this unique community known as The Woodlands, he may have been most proud that he had kept so many of the trees. Some twenty years earlier, when they were constructing the first roads in this 28,000-acre development thirty miles north of Houston, he, the boss, had made a nuisance of himself by stopping the work crews and reminding them to leave as many trees as possible. People loved to live among the trees. *He* loved to live among the trees, a passion that grew from frequent hunting trips in his boyhood. It was, perhaps, an unusual affinity for someone born to a storm-ravaged barrier island, but George P. Mitchell was a man of incongruous passions.

A decade later, he would be branded not as a tree-loving environmentalist but as the "father of fracking." The title was inaccurate, but it stuck because what had started as a quixotic quest had, after years of perseverance, changed the face of global politics, reshaped the American energy industry, and lowered fuel costs for every US household. Not since John D. Rockefeller had one single individual in the energy business made a greater public impact.

In his later years, as the influence of the OPEC cartel appeared to be waning and the United States began asserting itself as an energy powerhouse once again, George would take pride in the changes he had unleashed. In terms of his impact on the world, fracking would be his greatest achievement, but George didn't want to be remembered as an oilman. He suggested to his wife, Cynthia, that his headstone should describe him as "husband, father, ecologist, founder of The Woodlands."[1] Even after fracking reshaped the energy world, The Woodlands held his heart.

Far more than just a real estate development, The Woodlands redefined the American suburb when it opened in 1974, and it would bring ideas about sustainable development into the mainstream. Every detail, from how parking lots were positioned to how rainwater ran off each yard, was planned with an eye toward environmental repercussions—and, of course, leaving as many of his beloved trees as possible. The Woodlands became a laboratory for combating urban decay, careless development, and other problems facing American cities. Its ideas of sustainability would be copied worldwide.

How could a man responsible for what many environmentalists consider the greatest scourge of our generation also hold such a steadfast commitment to sustainability? George's son Todd once called this quandary "the Mitchell paradox," but George never saw the contradiction. His views did not fit neatly with those of his fellow oilmen or developers. Environmentalists who in the early 2000s would criticize him for ravaging the planet were largely unaware that in the 1970s he had been criticized for trying to save it.

As George began revealing his ideas for The Woodlands, some local developers laughed. Other seethed. This wasn't how things were done. What did an oilman know about development? "There was venom about George Mitchell," said Jim Blackburn, a Houston environmental lawyer who worked with George for years. "He wasn't a member of the gang, and he was doing something totally different. He was doing things that none of them did."

Houston, even in the 1960s, was a paragon of urban sprawl. City leaders showed less interest in preserving history than in encouraging wanton growth. Paving was progress. The price was borne by the natural landscape on which the city was built—a flat coastal plain that slopes gradually toward the Gulf of Mexico, entwined by the meandering flow of some 2,500 miles of bayous and dotted with pine forests and wetlands.[2] Founded as a business venture in 1837 by two New York City real estate promoters, the brothers John Kirby Allen and Augustus Chapman Allen, Houston was a city that didn't look back.[3] It became a center for trade, then energy, and by the 1960s, it welcomed the National Aeronautics and Space Administration's Mission Control, the Astrodome, and all things modern and futuristic—Space City, USA.

That was decades before George had perfected fracking, but he had already achieved success in the oil business. He was living with his wife and ten children in a 12,500-square-foot custom-built home in a tony neighborhood between Memorial Drive and Buffalo Bayou.[4] But even there, in one of the most modern developments in one of America's most modern cities, George found the surroundings lacking. He hated the way development was done in Houston—lots shorn of trees, bayou banks poured over with concrete, main roads clogged by strip shopping centers, parks and green space set aside with, at best, grudging indifference. He believed it could be done better, and he decided to prove it. "George Mitchell made an incredible statement with The Woodlands, and that statement was doing a green project in Houston," Blackburn said.

George, in fact, specialized in doing things no one else did—in energy, in real estate development, in historic preservation, and later, in philanthropy. In every industry in which he operated, he brought a lack of convention and a disregard for the status quo. His politics were more liberal than those of most of his oil industry counterparts, and his intellectual curiosity extended far beyond both energy and the environment, to physics, sociology, medicine, and education.

Success, just like fracking, came gradually for George. Quiet and unassuming, he represented a new breed of Texas wealth. The son of an illiterate immigrant who pressed clothes for a living, George was not the scion of an oil fortune like many of his better-known contemporaries such as the Hunts, the Basses, and the Waggoners. In fact, he was the opposite of the stereotypical swaggering Texas oil millionaire. He was unpretentious and even shy in social settings, and he shunned the more ostentatious trappings of oil wealth. He didn't have a private jet. When he flew, he insisted on sitting in coach. Rather than a yacht, he had a thirty-two-foot fishing boat, *Where the Fish Are*, that he used around his beloved Galveston, never straying too far into the open waters because he was prone to seasickness. What he lacked in braggadocio he made up for in soft-spoken intellect. From the time he was a young boy, George was fascinated by life's biggest questions. He pondered a career in astrophysics, but a childhood of financial hardship drew him to the potential profits of the oil business.

His fortune in energy would come less from oil than from natural gas, a clear, odorless hydrocarbon that was once considered a by-product of oil drilling. In many oil fields, natural gas is still burned off as waste in a process known as flaring. But over the years, it had also been used as a heating fuel, and although gas sold for less than oil, producing it could still be a profitable business. George preferred the fuel because there was less competition for it and because it burned cleaner than oil. As early as the 1960s, he worried that the world, and the United States in particular, was becoming too oil dependent. He feared that pollution would eventually destroy the planet. He saw gas as a cleaner alternative, a fuel for the future, at a time when most in the energy business considered it little more than a nuisance.

George demonstrated a rare ability to spot natural gas deposits overlooked by others. Unlike his peers, though, he didn't use his skills to spread his company's footprint across the country. He tended to find one deposit and focus on it. As a result, Mitchell Energy and Development, which he had founded in 1946 with his brother, never wandered too far from its native Texas. It never drilled overseas, and it showed little interest in offshore projects. But it fully exploited the fields it had. The crown jewel was some three hundred thousand acres to which it had leased mineral rights in Wise County, Texas, about forty miles northwest of Fort Worth.

That was where the process known as fracking became a common industry practice. For seventeen years, George insisted—and even demanded—that his engineers find a way to unlock the natural gas that he knew was trapped in the shale formations under his already prolific field. Petroleum geologists refer to shale as “source rock” because oil and natural gas form there and then migrate to more permeable geologic strata—limestones and sandstones—where they can become trapped in the pores of the rock to form what is known in the oil and gas business as reservoirs. No one, including Mitchell Energy employees, believed that gas could be extracted directly from shale—the rock was too solid. The hydrocarbons didn’t pool; they were held in the rock like water in a hard, dense sponge.

George was determined to unleash these shale-bound hydrocarbons because he saw that his company would soon deplete the natural gas in the Wise County field. He needed more to keep supplying his biggest customer, a Chicago utility that Mitchell Energy had locked into a long-term contract back in the 1950s. The utility had agreed to a price far above the market—an exceptionally lucrative arrangement. Mitchell Energy’s economic necessity, combined with years of trial and error, created a fracking technique that would be copied by rivals and go on to return the United States to a position of strength in the global energy market that it hadn’t seen in more than four decades.

Neither George nor his engineers invented fracking. The technique had been used since the 1940s to widen natural fracture networks inside rock strata and increase oil and gas production. But Mitchell Energy’s team found the right recipe of water, sand, and pressure to unlock those hidden gas deposits in the Barnett Shale formation buried thousands of feet under the gently rolling Wise County pastureland. It was as if they had discovered a second gas field under the first one. The new method, dubbed a “slick-water frack,” was unconventional. Engineers had long believed that gels, not water, were the only effective transport fluid for the sand that would keep the tiny fractures in the shale open long enough for the gas to escape.

The development of economically viable fracking took seventeen years of experimentation, much like a chef testing different ingredients before finding the right combination. The engineer who made the discovery spent months monitoring the gas production of the well before he became convinced the new fracking technique was successful. Even then, other companies dismissed fracking as

"Mitchell's folly." Its true benefits weren't fully understood until after George sold his company to Devon Energy in 2002. The expansion of fracking was gradual, and so was the realization that it was changing the production picture for the entire country. "I've been asked, 'When was the eureka! moment when you rushed out of the office?'" said Larry Nichols, who was chair of Devon when it bought Mitchell Energy. "I'm not even sure there was a eureka! year. It just gradually evolved."

George is often credited with using fracking to unlock the "shale revolution." But fracking alone didn't make America one of the world's largest natural gas producers. Wells that were drilled vertically into shale and then fracked did not produce enough gas to start a revolution. It took a combination of fracking and horizontal drilling, in which the well is drilled to a certain depth and then angled to cut horizontally across the shale formation, to yield enough gas per well to make fracking profitable.

At first, years of fracking produced only modest changes in Barnett Shale production. Only in the late 1990s, after Mitchell's team had tweaked the recipe of sand, water, chemicals, and pressure, did production begin to rise. By mid-2007 it had soared to 350 million cubic feet a day gained from 8,000 wells. As companies adopted Mitchell's techniques, natural gas production nationwide followed a similar trend. In 2015, the United States produced almost 50 percent more gas—27 trillion cubic feet—than it had in 2005. Back then, the country had braced for a gas shortage, and companies invested billions of dollars in terminals along the Gulf Coast that would import gas in liquefied form as domestic production waned.[5] A decade later, as fracking became widespread, America had more domestic gas than it knew what to do with. The companies that had spent billions on import terminals now spent billions more to convert them for exporting liquefied natural gas to countries in Asia and Europe. Gas prices plunged based on the abundant domestic production, and power companies began shuttering generation plants that used coal, which was now more expensive. In April 2015, the amount of electricity generated with natural gas surpassed that with coal on a national basis for the first time, contributing to a dramatic slowing of US carbon emissions.[6]

Other companies took Mitchell Energy's method for natural gas production and adapted it to oil, sending a shock wave through the world of petro-politics. In mid-2014, oil prices topped one hundred dollars a barrel. Then, as American companies pumped increasing volumes of oil from shale formations, prices began to fall. By the end of the year, they would slip below sixty dollars.[7] US imports of foreign oil fell to their lowest in two decades. Fracking unlocked oil in shale formations from North Dakota to South Texas and transformed the United States into a major oil producer for the first time since the 1960s. By 2015, fracking was saving every US household an estimated $2,000 a year in energy costs, adding more than $500 billion to the national economy, and creating almost four million jobs.[8]

Meanwhile, the United States quietly saw renewable energy production increase. Here, too, natural gas from fracking played a critical role. For all their environmental benefits, renewable energies such as solar and wind are not steady sources of power. They rise and fall with the sun and the breezes. This unpredictability makes it difficult—and expensive—for utilities to keep the lights on. Natural gas offers a solution. Older power plants that run on coal or nuclear energy can take days to adjust their output in response to changes in demand. Modern gas-fired power plants, however, can turn on and off quickly, reacting more rapidly and cost effectively to the intermittent generation from renewables.

As a result, the abundant cheap natural gas helped overcome the cost of renewables, which had impeded their growth for decades, expanding the use of economically viable "green" energy. Increasing generation from renewables and gas-fired generation cut the country's carbon emissions, and by 2012 the nation had complied with the Kyoto Protocol that President George W. Bush had refused to adopt in 2005 out of concern for the economic impact.[9] At the time, neither Bush nor the country's top energy advisers could anticipate the benefits of fracking, which had barely moved beyond George's original leases in Wise County.

Over the years, more utilities shuttered coal plants, enabling the United States to meet the terms of the agreement it never signed while avoiding the economic consequences that Bush feared when the country was still beholden to foreign oil imports. And the advance continued. In 2016, new installations of renewable energy surpassed those of conventional sources for the first time.[10]

All these achievements trace back to George and his team of engineers and geologists who pursued fracking for almost two decades, persevering through setbacks and failure. George poured millions into the project from 1981 to 1998, straining his company's finances, before the idea reached fruition. Facing his company's financial struggles in the 1990s, George, well past retirement age by then, could have sold his energy company and walked away. Had he done so, it's unlikely the buyer would have continued to invest in Mitchell's folly. It made no sense to keep backing a project that had failed to prove itself after years of trying. Someone might have bought George's company but probably wouldn't have bought into his vision.

Fracking made George a billionaire because he persisted until the technique yielded results. He sold his company just as natural gas prices were rising, which drove up the value of Mitchell Energy's newfound reserves. The wealth allowed him to live his final years as a philanthropist, giving millions to support varied endeavors, from the world's largest telescope to cancer research.

But for all his success, he was plagued by one major regret: the money came too late to save his dream of a "livable forest." Instead, he reluctantly opted to sell The Woodlands in 1997 to preserve his energy assets. From a business standpoint, the deal made sense. Real estate had long been a drag on Mitchell Energy's returns. But it was not what he wanted to do.

George viewed every project in terms of its potential. In the 1950s, when he first looked at developing a village on Aspen Mountain in Colorado, he saw not just the stunning natural beauty, but a vision of how homes and condos could cluster in harmony around its base. For George, development and beauty were complementary. Despite his love of trees, he helped finance his initial land purchases for The Woodlands from a timber company by continuing to log some of the acreage. He wasn't clear-cutting, a practice he abhorred, but he was still harvesting trees from his livable forest.

By the time of the sale, the development had become an even greater contradiction. George hoped to solve the problems of inner-city decay by creating his own suburb that supported green space and reduced light pollution. "The whole idea from the beginning was to create an alternative to urban sprawl that was actually feasible," said Robert Heineman, who has worked for The Woodlands Development Corporation since the early 1970s. Yet George's creation also drew Houston's blanket of concrete and asphalt northward, extending the sprawl of what was already one of the country's most spread-out cities.

The Woodlands inspired master-planned communities nationwide, yet in region after region these imitators, too, augmented the sprawl George sought to combat. Denton, a city on the eastern edge of the Barnett Shale where fracking was born, voted to ban the drilling technique inside its limits in 2014. Just a few miles away, within sight of the first well Mitchell Energy drilled into the Barnett, homes are creeping ever closer to the wells. The neighborhoods advancing toward the gas field are planned communities that in some ways took inspiration from The Woodlands. There, on the plains of North Texas, the Mitchell paradox is a physical clash as well as a philosophical one.

George wasn't thinking of any of those things, though, as he sulked past reporters and photographers on that Friday in 1997 after announcing the sale of The Woodlands. Normally he was an optimist, but on this day, he was consumed by an uncharacteristic sadness. The droopiness in his stride seemed to shave a couple of inches from his five-foot-nine frame. When had he felt like this before? Perhaps not since his mother's sudden death when he was thirteen years old. "After 31 years of effort, it's kind of sad to say you won't have much to do with it anymore," George, watery eyed, told a reporter from the *Houston Chronicle*.[11]

On that morning, fracking was still just a desperate idea, and George's entire universe seemed to be crumbling. The Woodlands was gone, his energy company was struggling, and health issues loomed for him and his wife, Cynthia. As he walked to his car in the morning heat, he was stoic. The quiet confidence with which he had built his company seemed to falter. In his right hand, he held a few papers from which he had read moments earlier during a press conference that had felt more like a wake. There were no handshakes, no pretend smiles, little more than a nod to the reporters with whom he was typically happy to speak. He'd put on a good face. He'd made it sound as if everything had been his idea, a strategic business decision, something that had to be done. All of that was true, of course, but George didn't make business decisions in the same cold, calculating way other businessmen did. For him, they were often instinctive or emotional, and few had ever been more emotional than the sale he'd just announced.

Spiros Vassilakis, who had worked for George for almost a decade at that point, had never seen his boss so dejected. "When I saw him walking out of that room, he was not a happy man," Vassilakis said.

At the time, Mitchell Energy and Development, the company George had cofounded in 1946 when he was twenty-seven years old, had annual revenue of $1.1 billion and 1,950 employees.[12] Its oil and gas properties blanketed the plains west of Fort Worth and dotted fields in East Texas and along the Gulf Coast. The company collected $543 million from the sale of The Woodlands, and despite George's personal attachment to the development, his investors were glad to see it go.[13] Mitchell Energy's stock soared on the news, and George, as the company's biggest shareholder, saw his personal wealth increase. But for him, The Woodlands had never been about the money, and his company had never been solely about pleasing shareholders. The Woodlands was a chance to improve how people lived and worked in America's cities. It was something he had dreamed about, had raised from soggy pine forests, had fashioned and fretted over and nurtured. It wasn't just a development; it was his home and the home of many of his employees, all of whom viewed it as the physical embodiment of George's vision.

Had it been up to him, he would have sold his company's oil and natural gas business, but he'd had little choice. His board of directors, which included his son Todd, had been pressing him to split the company's energy and real estate holdings. Many investors balked at George's pet projects that siphoned profits from the energy business. Selling the oil company, as much as he would have preferred it, was impossible at the time. The company was embroiled in a long-running lawsuit with landowners in areas near his North Texas gas field. Who would buy the firm with the uncertainty of legal liability dangling like a sword over the company's future? Reluctantly, he came to the only logical decision—he would have to sell the community he thought of as his eleventh child. The following month, in July 1997, his public relations department sent him a copy of a brief press release noting that the deal had closed. The sale was final. In approving the wording, he wrote three words at the top of the draft: "OK but sad."[14]

If George's vision for fracking was born of necessity, his vision for The Woodlands was born of love. By the time of the sale, The Woodlands had grown from uninhabited timberland to a city of about 50,000 people.[15] He was proud of what it had become, and he believed it had yet to reach its true potential. After all, he hadn't made a dime on the project. Like fracking, his sustainable development endeavor was a financial sinkhole, and he kept pouring money in only because he retained a controlling ownership stake in Mitchell Energy. Future owners might follow his plan, but their profit motive would become evident in every aspect of the development. Residential lots got smaller, stands of trees got thinner, buildings got taller.

The Woodlands was well established, but the work wasn't finished. Its mission had not been fully accomplished. This made the sale even more shocking to the employees who worked on the project. "We were all disappointed," Heineman said. Other developers have studied The Woodlands, and many communities boast of their planning, but few are willing to take the financial risk that George did. Despite the sale price, the company's investment far outweighed its return. George was willing to wait decades for profits—far longer than most other developers. "To do a project this large today is almost impossible," Heineman said.

George put more of himself into The Woodlands than anything he'd ever done in the oil field. Reflecting on The Woodlands sale in 2004, before fracking had reached its full effect on global energy markets, he said: "I hated to sell The Woodlands, not only because it's more visible, but it probably has a greater effect on (people) compared to what I accomplished in energy." Rarely has anyone been more wrong in assessing his own legacy.

CHAPTER 2

The Greek with the Irish Name

Savvas Paraskevopoulos was leaving for good. Though he was only twenty years old, he could see his future and it wasn't a pretty picture. Born in February 1880, Savvas grew up in Nestani, a small mountain village in the Peloponnese peninsula, the southernmost region of the Greek mainland.[1] Steeped in legend, the Peloponnese is home to Sparta, one of the most famous cities in Greek history and the site of the first Olympic Games in 776 BC. The mountains often have snow dusting their bare peaks, and the valleys below are lush and lined with cypress trees, citrus groves, and vineyards.[2]

At the time that Savvas decided to leave, though, the Peloponnese was also among the poorest regions in the country. His parents, Christos Paraskevas Paraskevopoulos and Angeliki Revelioti, had married around 1860.[3] Greece had gained its independence less than forty years earlier, breaking free of the crumbling Ottoman Empire, and independence only increased the economic isolation of the Peloponnese.[4] In the 1890s, Greece endured a major crop failure and an economic depression and struggled with an unstable government. In addition, marriage depended on a dowry system that required fathers to provide lavish gifts such as homes and land when their daughters wed, which placed an economic strain on a rural economy dominated by agriculture.[5] As Savvas neared adulthood, many rural residents were moving to the cities in search of a better living. The Paraskevopouloses, though, stayed behind, engaging in one of the most traditional Greek enterprises—raising goats. The family of eight lived in a small rock hovel with a dirt floor.[6] As a young boy, Savvas led the family's small herd to graze in a patch of pastureland his family owned near Milia. The land was too small, though, to do

Savvas much good in adulthood. He was the youngest of six children, including older sisters. Even if their dowries didn't consume most of the land, his older brothers would get their shares first. He might wind up with half an acre at most—not enough property on which to make a living.[7] He had little hope of buying more with what he could make on a goatherd's pay.

He set out on foot for the port city of Kalamata, on the southern tip of the peninsula, where the Greek fight for independence began in 1821. Now, as a new century dawned, Savvas found passage on a ship that would take him to a land that promised everyone a chance for a fresh start.

He arrived in America following the well-worn path of millions of immigrants, sailing under the raised torch of the Statue of Liberty and through the entry point of Ellis Island.[8] Almost immediately, he set about finding work. As an unskilled laborer—America didn't need many goatherds by the late nineteenth century—his job prospects were limited. The railroads had been expanding their tracks across the country since the end of the Civil War, and they were hiring large numbers of immigrants for manual work. Fortunately for Savvas, they didn't care if the men they hired couldn't speak English. Savvas not only didn't know English, he couldn't read his native Greek.

A week after his arrival, he was working on a rail gang in Arkansas, laying track for thirty-seven cents a day.[9] Almost immediately, young Savvas ran afoul of the short-tempered Irish paymaster. "The paymaster one day he come to me and say 'listen, I can't say your name; I can't spell your name, and if you don't do something about it, I'm going to fire you,'" Savvas recalled in a 1963 newspaper interview. "I ask him what's his name, he say 'Mitchell,' and I tell him 'I'm Mike Mitchell,' and I was from that day to this."[10]

A new name didn't change the fact that he was still a young man in a strange land, far from home and family. One day a foreman, who was also Greek, found the young Savvas sobbing. Savvas said he wanted to write his mother to let her know he was alive and safe, but he didn't know how. The foreman spent the next year teaching Savvas to read and write in his native Greek.[11] Finally, he could communicate with his family, although his command of English would remain rudimentary for years.

The newly minted Mike Mitchell followed railroad work to Utah and after a few years made his way to Houston, which had become home to a growing community of Greek immigrants.[12] In 1860, the census in Texas recorded only two Greeks in the entire state—an area five times the size of Greece. Over the next fifty years, however, that would change. By 1910, one-tenth of Greece's population had come to America, and in Texas they flocked to the major cities, especially Houston. It was close to the coast, and the hot, humid climate had the feel of home. By the time Mike arrived 1905, Houston and the island city of Galveston farther south both had established Greek Orthodox churches, which became a rallying point for attracting more Greeks to the cities' burgeoning immigrant communities.[13]

Among those drawn to the area were Mike's cousins Alexander, Andrew, and Jim Chiacos, who had also emigrated from Nestani a few years earlier. After four years working on the railroad, Mike was ready for a different line of work, and the cousins had established a shoeshine parlor and barbershop in downtown Houston. Mike joined them, and the business became known as Chiacos & Mitchell. Money was tight in the early years, and Mike soon fell into arrears at a local mercantile, Haubert Bros., for $165. The merchant took him to court, and the judge ordered Mike to repay the store with interest.[14] Gradually, though, Chiacos & Mitchell did well enough that the cousins expanded into the clothes-pressing business, which was housed on the second floor of a building they leased on Main Street for $625 a month.[15] Eventually, they opened an eatery, People's Restaurant, which offered delivery service.[16] Mike, who was prone to hyperbole, recalled success coming more quickly: "I had one shine chair when I started, but in six months, I have 10 chairs!"[17] Success may been more muted, but Mike worked hard. He developed a double hernia that would bother him for most of his life and wasn't corrected until he was seventy years old.[18]

Less than a decade after leaving Greece, the former Savvas Paraskevopoulos had a new name, a new country, and new life as a businessman in one of America's fastest-growing cities. As he established himself, his outgoing nature helped him make friends among doctors, lawyers, and other members of the city's social elite. Although he had no formal schooling, Mike was smart and had a remarkable memory. His oldest son, Christie, would later claim that Mike could remember ten thousand phone numbers.[19] The young man who had cried because he couldn't write home now exuded self-confidence and was popular in social settings, particularly among Houston's Greeks. He was a snappy dresser who claimed to own fifty suits that he changed as often as three times a day. As he approached age thirty, he decided he should get married. His method for finding a wife, however, was unconventional.

Katina Eleftheriou, if you believe the Mitchell family legend, was a modern-day Helen of Troy. Born in 1888 in Argos, which is also in the Peloponnese, she, too, came through Ellis Island not long after the railroad paymaster in Arkansas rechristened Savvas Paraskevopoulos as Mike Mitchell. Family members tell different stories of how Mike and Katina met, and not surprisingly, Mike's is the most lavish.

In what has become known in the Mitchell family as "the newspaper story," Mike claimed that Katina was so beautiful that her arrival in the New World generated a story in a Greek newspaper (some accounts say it was a magazine) that included her photo under the headline "Greek Beauty Arrives in the US." "I frame the picture and hang it in my bedroom and I say I will marry Katina," Mike recalled. "I spend three years and $13,000 looking for Katina." He finally tracked her to her sister's house in Tarpon Springs, Florida. "Already, two millionaires are proposing to her, but she dropped 'em for me. I was a handsome man. I had good suits and I wear a flower in my lapel every day and I changed my clothes three times a day. Katina, she marry me."[20] In other versions of the story, Mike had to overcome only one millionaire, but Katina was already engaged. He had to convince Katina's brothers to let him speak with her. He won over both the brothers, and then, worn down by his dogged pursuit, Katina broke off her engagement and agreed to return with Mike to Texas.[21]

Mike told the story so often that it shows up as fact in newspaper articles and books about George Mitchell.[22] George himself frequently told variations of it, and by then, Katina wasn't around to refute it. George's brother Christie, a columnist for the *Galveston Daily News*, further embellished the story in a 1970 column.

According to him, Mike didn't win Katina's hand on his trip to Florida. Instead, she came to Houston, and a local preacher tried to persuade her to marry a prominent Greek restaurateur. The would-be suitor threw a lavish party in Katina's honor, to which Mike wasn't invited. He crashed the party armed with the biggest revolver he could find. "Mike walked in on this party like Jesse James, flashing his big six-shooter, and said: 'Katina, get up and come to my side. And if you Greek traitors love life on this earth, you will kindly keep your seats until we depart.'"[23]

The reality was probably far more mundane. Katina did, indeed, emigrate from Greece in June 1910 with her family, coming through Ellis Island and settling in Tarpon Springs, which, like Houston, was a coastal community with a growing Greek population.[24] Many Greeks who made their living as sponge divers found bountiful harvests off the western Florida coast, and today the city is known as the "Sponge Capital of the World."

Katina, however, also had relatives in Houston, and it's likely that Mike met her when she was visiting them. Mike traveled to Florida and, as was the custom in Greek culture, convinced her brothers that he could support her. He did, after all, have a successful business in Houston, which probably made him more appealing economically than many other Greek suitors who were less established in America. Mike's natural gift for self-promotion no doubt enhanced his portrayal of Chiacos & Mitchell's success. If Katina were, indeed, engaged to someone else, Mike probably presented a better bargain to the family.[25]

Regardless of how Mike won Katina's hand, marriage records show they wed in Texas on May 27, 1911, and moved south of Houston, to the island city of Galveston.[26] They were a handsome couple. Mike, ever the sharp dresser, had a thin mustache fashionable at the time. His hairline was in full retreat even though he was only thirty-one years old, giving him an air of maturity and sophistication. Katina had a gentle, doll-like appearance, and she looked much younger than her age, which was twenty-three when she married Mike. She had long, dark hair and gray-green eyes.[27] In pictures, she conveys the sense of gentleness for which she would be remembered by her children. After a few years on the island, she had become a leader in local Greek society, known for her amiable and gentle disposition, as well as for being "one of the most beautiful Greek ladies in the South."[28] Soon after their wedding, Mike gave Katina a phonograph on which they would play records of Greek music. Mike would keep the phonograph the rest of his life, reminding the nurses who took care of him in his final years that he'd given it to his beautiful bride.[29]

On the island, Mike opened a Chiacos & Mitchell location on Market Street, in the Strand District—the side of the island facing the mainland. Operating under the name "Neat Dressers' Pressing Club," the shop advertised shoe shining services for men and women that Mike touted as "the best in the city." Eventually, he also opened a "palatial" barbershop nearby.[30]

Galveston was one of the oldest cities in Texas, and one of its most vibrant. In the 1800s it developed a thriving port, with the island itself serving as a breakwater for ships docking near the Strand. Its streets bustled with commerce fueled by a railroad that spanned the channel to the mainland and brought a steady supply of cotton to be shipped overseas. By the time Mike and Katina arrived, though, Galveston was rebuilding itself. The city had been leveled a decade earlier, in 1900, by one of the worst hurricanes in American history.

The storm roared across the Gulf of Mexico and slammed into the unsuspecting island with winds of 145 miles per hour. A fifteen-foot storm surge washed over the shoreline, which sat less than nine feet above sea level. Officially, the death toll was listed as eight thousand, but thousands more bodies were never recovered. It remains the country's deadliest natural disaster.

The Mitchells arrived in the midst of Galveston's rebirth, though signs of the devastation remained. The island had almost no large trees.[31] In 1904, the city had unveiled plans for a seventeen-foot-high seawall, backed by granite boulders that would protect it from another storm. *McClure's* magazine declared it "one of the greatest engineering works of modern times." But city leaders didn't stop there. They raised the center of town—more than two thousand buildings in all—with manually cranked screw jacks and filled the gap underneath with some eleven million pounds of dirt. The colossal task was completed just as Mike Mitchell and his new bride arrived in town.[32]

Galveston emerged bigger, wealthier, and more vibrant than

before the storm. For a while, the port would return to its previous level of activity. But the barrier island couldn't shake its image of vulnerability. Congress had approved the dredging of the Houston Ship Channel in the late 1800s, and major shipping activity would gradually migrate up Galveston Bay toward Houston. A century later, Houston's port would grow into one of the country's busiest and become an engine for growth while Galveston endured decades of economic stagnation.

The Mitchells were just one family among many drawn to the island. To fuel its rebirth, city leaders welcomed immigrants from all over the United States, many of whom, like Mike, had arrived in the early years of the new century and were still looking for the financial opportunity on which to build their American Dream. Galveston's neighborhoods were already integrated by then, with black families living on the same blocks as white ones. Segregation persisted, of course, but Jim Crow laws were loosely enforced or even ignored. Black children frequently played on the Seawall, mixing with the island's stew of other races and nationalities. Local immigrants, especially members of the growing Greek population, helped Mike and Katina assimilate. At the time, Mike still spoke little English, and Katina would speak only Greek for the rest of her life.

Home to the state's only two Greeks in 1860, Galveston, with its bustling port and thriving commerce, had become a lure for a growing Greek community by the time Mike and Katina arrived. Initially, many Greeks were drawn to the city's seafaring culture, working as fishermen, sailors, and merchants. But as their numbers grew, Greek immigrants became saloon owners, grocers, and cotton gin operators, and they mingled with the Serbians, Russians, Syrians, Italians, and other ethnic groups that coalesced in the island city.[33]

Mike and Katina's life together was a constant struggle for money, but Mike always managed to scrape together enough for the family to get by. "He played his wits," George would recall. "He had a very sharp mind, completely untrained [and] he had tremendous perseverance. He was friends with everybody in town—all the doctors and all the lawyers. They all did things for him for nothing."[34]

George may have understood his father better than anyone, appreciating Mike's intelligence and tenacity, but Mike's oldest son,

Christie, summed up his father's enduring qualities in a letter to his sister: "If the old man had not had a sense of beauty, the determination of a Georgia mule and the brains and ingenuity to get what he wanted, none of us would be here."[35]

In addition to the Mitchells, members of another immigrant family, the Maceos, arrived from Sicily about the same time as Mike and Katina. One of the Maceos would play a crucial role in the future of Mike Mitchell's youngest son—and ensure that the Mitchell name endured on the island long after the Maceos were no longer household names.

◈

On a late summer Sunday morning in 1914, Mike got a phone call from one of his cousins. A fire had broken out at the Seward Building, which housed Chiacos & Mitchell's main shop in Houston. A passing newsboy had noticed smoke pouring from the windows and notified the fire department. As firefighters arrived, an explosion rocked the building, which shared walls with restaurants on either side. Moments before the fire broke out, four men had sat down at the counter of the White Kitchen next door to order breakfast. As the explosion ripped through the common wall between the two establishments, the roof of the White Kitchen collapsed on the men. Two of them died in the blast and the other two were taken to a nearby hospital with spinal injuries. Both were paralyzed and later died—one of them after battling paralysis and other trauma for two months. About twenty other patrons in the White Kitchen and a restaurant on the other side of the Seward Building were also injured, as well as two police officers who responded to the disaster.[36]

"I was sitting about 10 feet from the door, and suddenly there came the explosion," Earl Eagan, one of the men seated at the counter, told the *Houston Post* from his hospital bed. "Everything went black and I was rendered unconscious. I could feel the water striking me that the firemen were [spraying] on the building, but my body seemed numb all over. Finally, after what seemed an age, someone got to me and I was pulled out." Eagan, one of the two paralyzed men, died a few weeks later.

L. E. Ogilvie, a policeman who was patrolling nearby at the time of the explosions, threw off his coat and rushed into the rubble. "We could hear men under the mass of timber and brick calling to us for help, begging for someone to come to them," he said. "The first man we got to had his hands folded. A little further on down, I felt a man on the fourth stool. He was dead. We didn't bother with him but went on after a man who was calling to us."

Ogilvie, a former fireman, rescued several of the patrons that day. By noon, firefighters and the police were investigating a possible cause—a can and several jars of gasoline stored in the Chiacos & Mitchell pressing shop on the second floor. A city ordinance prohibited storing flammable liquids above the ground floor in commercial buildings.[37] By noon, the fire marshal had issued warrants for Mike and his cousins.

After hearing about the fire, Mike caught the next train into the city from Galveston, and when he arrived, the police promptly arrested him on charges of arson and murder. He was taken to police headquarters with his cousin Alex. The police questioned the two men for more than an hour before jailing them.[38]

At a bond hearing, which had to be moved to the courthouse basement because there were too many spectators to fit in the regular courtroom, the night cashier at one of the two devastated restaurants testified that he had seen Jim and Alex Chiacos walking past the building at 2:00 a.m. on the morning of the fire. A printer whose shop shared part of the building's ground floor with Chiacos & Mitchell claimed he'd heard "unusual noises" from the second level before the fire. Several insurance company representatives testified at the hearing, including one who said that Mike had bought personal policies on the building and his company's stock worth a total of $5,500 just weeks before the fire.[39] The implication was clear: the fire marshal believed that Mike and his cousins had torched their shop to collect the insurance money.

However, a bootblack who worked for the cousins testified that Jim and Alex Chiacos had walked him home at 12:45 a.m. Mike maintained that he was in Galveston at the time of the fire. Prosecutors argued that Chiacos & Mitchell was struggling financially, but they failed to produce much evidence to support the claim beyond a few small unpaid bills. Attorneys for Mike and his cousins suggested that faulty wiring might have been the cause of the blaze, although that didn't explain the two explosions that ripped the building apart. The judge set bond at $2,000 each, which Mike managed to post and get himself released.[40]

On November 2, a day after the fourth victim died, the Chiacos brothers were indicted on charges of arson and murder, and Mike was charged with two counts of being an accomplice to arson.[41] The cousins surrendered and spent five nights in jail before posting bond.[42] Despite the severity of the charges, the ordeal concluded with little fanfare just before year's end, when a judge dismissed all charges.[43]

Perhaps Mike used his contacts in the legal community to get the charges dropped—he always boasted about the number of judges and lawyers he knew—or perhaps there wasn't enough evidence to convict him.[44] Had Mike and his cousins set fire to their own business? Mike's whereabouts at the time of the fire were never confirmed, and a family friend would later say that Mike wasn't in Galveston when the blaze started, despite what he told the papers and the police.[45]

Regardless, Chiacos & Mitchell couldn't survive the financial fallout from the fire. Even with Mike's policies, the insurance wasn't enough to cover the losses, and the partnership disbanded. The Seward Building was rebuilt the following spring and a new tenant moved into the Chiacos & Mitchell space. Mike, meanwhile, retreated to Galveston and went into business for himself.[46]

CHAPTER 3

George

Mike managed to hang on to the old Chiacos & Mitchell shop in Galveston's Strand District, renaming it the Neat Dresser's Club. He had a simple formula for success: "I shine shoes for a nickel—five cents—and I clean and press suits for 50 cents. When the others go up to $1, I still press for 50 cents."[1] He was becoming increasingly well known on the island. He and Katina were living in an apartment over the business, and two years before the Houston fire, Katina gave birth to their first child. They named the boy "Christ," though everyone would know him as "Christie."[2]

The same year as the fire, Katina had another son, Johnny.[3] In Greek culture, daughters came with the expectation of dowries, which could be financially crippling for families. Sons, on the other hand, had no such requirements, and even though the dowry system was dying out among immigrants in America, two sons still carried a certain status in the community, which only added to Mike's reputation. In 1916, Katina had another son, who died at birth.[4]

Despite the heartbreak of losing a child, the couple persevered. They became active in helping other Greek immigrants adjust to life in Galveston. Mike balanced this community outreach with his other favorite pastimes—drinking and gambling, especially cards. "He was an honest, lecherous Greek," their daughter, Maria, would recall. "He loved his family unconditionally, but he frequently stayed out late playing poker and chasing women."[5] Galveston was a rough-and-tumble place, and Mike found himself in more than one scrape. A fellow card player accused him of cheating, pulled out a gun, and fired. As Mike told the story, the bullet passed through his hat but missed his head.[6]

Katina, meanwhile, was in many ways the opposite of her husband—quiet, thoughtful, and soft-spoken. She was active in the Greek Society of Galveston and known throughout the island for her generosity and understanding. She was a good cook who, in addition to making meals for her family, often cooked for the steady stream of guests that Mike would bring home.[7]

On May 21, 1919, Katina gave birth to another son, George, and a year later, on Christmas Day, to a daughter, Maria. George was named by his godfather, a friend of Mike's who was visiting from New York. The man was the godfather of as many as a dozen children, and he named them all "George" or "Georgia."[8] The six-year gap between Johnny and George created a natural division among the siblings. They remained close for their entire lives—and George and Johnny would become business partners—but as the two youngest, George and Maria shared a special bond. They had their mother's quieter, more introspective nature, while the older boys had Mike's outgoing personality as well as his stocky build and dark, chiseled features. George was taller and thinner, with intense greenish-brown eyes. He wasn't as athletic as his older brothers. Next to them, he looked skinny—"so thin they called him 'shrimp.'"[9]

With four children to provide for, the family struggled. Mike's business barely made enough for them to get by, but he managed to rent a comfortable home on the east side of the island.[10] Galveston was among the more economically stable places in the country for a young immigrant family. Its drive to rebuild itself after the Great Storm resulted in a tight-knit community that relied on the port and the local military base, Fort Crockett, for most of its jobs. The rest came from shopkeepers like Mike and establishments along the Seawall that catered to tourists. World War I had ended a year before George was born, but the war had little impact on Galveston other than some disruptions to shipping, which affected the port. In 1921, when George was just two years old, the United States was entering a depression, but in Galveston most people had jobs, and grain exports broke records that year.[11]

Although Mike and Katina had little formal education between them, they pushed their children to do well in school. By the time he was four, George was clearly the most studious of the siblings. As he started school, Georgie, as his friends called him, spent hours in the library devouring books, such as the Tom Swift series, which was among his favorites.[12] By the time he was fourteen, he had read most of the books in Galveston's public library, and he was doing so well in school that he skipped two grades.[13]

He spent his early years in the local Greek school, which Christie and Johnny also attended. Classes were taught in Greek, which the Mitchells all spoke at home.[14]

After Katina married Mike, most of her family followed her to Galveston. Her brother, Jimmy (who changed his last name from Eleftheriou to Lampis after he arrived in America), taught the Mitchell boys to fish. Mike's family had been mountain goat-herds, living miles from the sea. Katina's family, though, hailed from closer to the coast, and "Uncle Jimmy" was a master fisher-man who passed on his skills to his nephew. George put his angling skills to work helping his mother. He would get up early in the morning, head to the beach and catch fish, and then bring them home to Katina before going to school. Frequently, the catch became the family's dinner that evening.[15]

Uncle Jimmy also taught George to hunt quail on the largely unin-habited expanse of Galveston's West Beach, where the native grasses created a perfect habitat for ground-nesting birds. On their first trip, Jimmy gave George a 12-gauge shotgun that had so much recoil it knocked the skinny boy to the ground every time he shot it.[16] Uncle Jimmy had a fondness for George, who had a calmer demeanor than his brothers and at an early age showed a head for figures. Jimmy ran a beachfront restaurant and bathhouse on the east end of the island, and while George was still in elementary school, Jimmy hired him to work the cash register because he could make change quickly and made fewer mistakes than some of the adult cashiers.[17] "I learned how to handle money real early," George recalled.[18]

He may not have had his brothers' outgoing nature, but George was every bit as enterprising. In addition to his work at his uncle's bathhouse, he would earn extra money by scouring the beaches for shells and sand dollars and then sell them to Murdoch's, a tourist shop on the Seawall.[19]

In 1929, when George was ten, the stock market crashed. He read in the *Galveston Daily News* about the travails on Wall Street and bank runs across the country. Texas was hit hard by the economic collapse. The Depression wiped out many wildcatters in the state's nascent oil business, and farmers—still the backbone of the state's economy—struggled. Corn that cost eighty cents a bushel to plant was selling for ten cents at harvest, and on the mainland, farmers were dumping their crops in the street. Galveston, however, managed to deflect some of the financial turmoil. No one rioted for lack of food, and because the island had no factories, there were no mass layoffs or demonstrations. None of the island's banks failed. The port, the city's major employer, saw a slight decline in business, but nothing that reflected the economic devastation going on elsewhere in the country. European nations continued to demand cotton, which flowed by rail from the mainland fields through Galveston's docks.[20]

But what really kept the Depression at bay was Galveston's shadow enterprises: bootlegging, gambling, and prostitution. The port had long attracted a transient element, and Galveston developed a reputation as an "open city," where vice was largely tolerated or at least overlooked. The Galveston of George's youth could be a wild, almost lawless place.

The year George was born, fast schooners known as "rum runners" darted from Cuba, Jamaica, and the Bahamas, carrying as many as twenty thousand cases of illicit liquor at a time. After Prohibition, the activity intensified. The schooners would anchor at "Rum Row," a point thirty-five miles off the coast, where the cargo would be transferred to smaller boats that could slip undetected onto the deserted beaches on the island's west end. From there, the bottles were wrapped in burlap sacks and stored in warehouses until they could be transported inland to cities as far away as Omaha and Denver. As Galveston became an increasingly active hub for illegal booze, it also became a haven for organized crime. Two rival gangs divided the island. Gun battles frequently erupted in the back streets and alleys. Frank "The Enforcer" Nitti reportedly got his start in the Galveston gangs before ripping off one of the leaders for $24,000 and fleeing to Chicago, where he became Al Capone's right-hand man.[21]

Mike's love of a good time, his affinity for cards, and his eye for women brought him into contact with some of the island's less savory elements. In 1929, having moved his business to Twenty-Third Street, closer to the Seawall, he opened a pressing shop and café known as Snug Harbor. In a crackdown on gambling by the county attorney that year, Mike was accused of allowing, and perhaps even encouraging, craps and poker games at his establishment.[22]

Meanwhile, the Maceo brothers, Sam and Rose, had become barbers and then expanded into an array of Galveston's tawdry businesses. Sam cut hair in the Galvez Hotel on the Seawall, and the Mitchell boys were regular customers.[23] Rose, meanwhile, had a single chair in the corner of a seafood canteen on Murdoch's pier, which was also a hangout for the rum-running Beach Gang. One of the leaders offered to cut Rose in on the action, and before long, Rose and his brother were partners in the Hollywood Dinner Club on the west edge of town. The club was the first in the country to have air conditioning, and it was adorned with crystal chandeliers and a huge dance floor.[24]

Later, the brothers would strike out on their own and open the Balinese Room, a notorious front for gambling and liquor sales. The Balinese would endure as an entertainment hot spot for decades after Prohibition, hosting Hollywood stars and celebrities.[25] For thirty years, the Maceos would run Galveston, at least behind the scenes. They put slot machines in all the restaurants—whether the owners wanted them or not—and they distilled booze in their own warehouse. The lawmen looked the other way because most were on the Maceos' payroll, as were the local judges and politicians. It wouldn't end until 1957, when the state's attorney general filed a civil lawsuit against the Maceos and their top lieutenants. To the locals, though, the Maceos—especially the affable Sam—were heroes who gave generously to churches and charities and people in need.[26]

There's no record of Mike Mitchell ever doing business with the Maceos, but he certainly knew them, and they may have formed a friendship over the poker tables. Given the Maceos' control of the island's gambling operations, it's likely Mike frequented some of the brothers' speakeasies.

◈

Mike Mitchell could get money far easier than he could keep it. Snug Harbor was a barbershop, shoeshine parlor, and café, but it also provided space for a little gambling on the side. Mike rented out rooms upstairs, but none of the ventures made much money, and the clothes-pressing machine was almost repossessed at least once.[27] Despite the family's financial struggles, Mike couldn't suppress his affinity for nice clothes, which he still changed frequently, and for playing cards. While he spent freely on himself, he also lavished gifts on his children, whom he loved unconditionally. He would take Maria shopping and on impulse ask her if she liked a particular pair of earrings. If she said yes, he would buy them for her.[28]

George always admired his father's uncanny ability to scrape together enough money to keep the Mitchells going.[29] He and his brothers inherited their father's enterprising skills, and although their ventures were less flamboyant, they were no less creative. In an early burst of ingenuity, George won the island's annual fishing roundup on a technicality. First place went to the person who caught the most fish, but the rules didn't specify the size. George got a net and scooped up dozens of finger mullets, tiny bait fish that proliferate in the shallow waters of the island's coves. He presented his haul to the judges, who apparently hadn't anticipated the loophole. They named George the winner. George didn't care about the trophy, but he wanted the rod and reel that came with first place. He would use that equipment to catch much bigger fish.[30] Several photos of George in his youth show him with catches that were larger than he was. In one, he's about fifteen years old and weighs about 110 pounds and is standing with a 120-pound tarpon he caught off a fishing pier on the north side of the island.[31] A year later, when he was sixteen, George got a mention in the local paper for landing a five-foot, six-inch tarpon—a fish so enormous that an adult had to hold it up for the photo.[32]

By the time George was twelve and his brothers were in high school, they had perfected a moneymaking scheme. They would hang out at Bettison's, a popular fishing pier, and catch batches of red fish and speckled trout. With their stringers bursting, they looked for tourists willing to buy the catch for two dollars apiece. Some of their customers were unlucky fishermen who wanted something to show for a day's fruitless outing, but others were men who came to the island and spent the weekend drinking, gambling, and whoring and needed to return home with fish to avoid raising their wives' suspicion. If no tourists were buying, the boys would sell their bounty to local restaurants such as Gaido's on Murdoch's pier.[33] After the tourists left for the summer, young George often supplemented his fishing income by clearing tables at the restaurant.[34]

One summer, he carefully set aside money from his fish sales, patiently building up enough cash for a new bicycle just in time for school. He rode it to class on the first day, bursting with pride. The bike was his first real investment, the first time he had planned and worked patiently toward a goal. Two days later, someone stole it. George never found out who did it, and he never found the bike, but decades later he still teared up telling the story.[35]

The boys also worked for Uncle Jimmy at the Lampis Bath House. "My job was to fold the dry towels and bring the wet towels up to the bath house," George recalled. "Johnny and Christie worked as lifeguards, chasing the girls."[36] Like their father, Christie and Johnny both made friends easily. George, on the other hand, preferred his own company. Katina decided to pull him out of the Greek school and enroll him in the public Sam Houston Elementary because she believed he would get a better education. Despite the family's financial struggles, she pushed all the kids academically and was determined that they all would go to college—even Maria, which George would later recall was unusual for a Greek daughter.[37] Sending four children to university would have been a challenge for any working-class family during the Depression, and Mike's spendthrift habits made it more difficult. Christie enrolled in Texas Agricultural and Mechanical College, as did Johnny two years later.

To save money and help pay for their schooling, the family moved from what had been a modest but comfortable home in the Strand District to three small rooms above the restaurant on Twenty-Third Street, about one block back from the beach.[38] The quarters were tight, because in addition to the four children,

Katina's ailing mother also came to live with them.[39] And Mike was still renting other rooms on the floor to help make extra money.

As the 1930s wore on, even Galveston's illicit economy couldn't shield the island from the ravages of the Depression. By September 1932, Christie had dropped out of A&M and was trying to make his way through the more expensive University of Texas at Austin. The family had run out of money, and it seemed that Christie would have to leave school permanently. On a Saturday night, two days before the registration deadline, Christie burst through the door of the apartment with good news: he had gotten a "scholarship" that would pay the rest of his tuition. The money may have come from Katina's brother, Jimmy Lampis, or it may have come from Sam Maceo—or both. Maceo was known to help Galveston's immigrant families when they fell on hard times. Regardless, it was enough money for one more year of college.[40] Katina's dream for her children to get degrees was still alive. Mother and son embraced and began to dance in the tiny kitchen. Suddenly, Katina collapsed in Christie's arms.

Even though she was only forty-four, she suffered from high blood pressure, and the excitement had triggered a stroke. She lapsed into a coma and never recovered, dying five days later.[41] The family was devastated by the loss. George, who was thirteen, took his mother's death particularly hard. Cesare Galli, a lifelong friend who attended Katina's funeral, remembered looking at George during the ceremony. "He was lost," Galli said.[42] Mike was too, and while he was still coping with Katina's death, he was hit by a car, breaking his leg in twenty-two places.[43] The combination of his injuries and the loss of his wife crippled Mike's business and left him unable to care for the younger children. Katina's family offered to help. Uncle Jimmy and his wife, Lela, had moved to Dickinson, just across the causeway on the mainland, and George went to live with them. Maria was sent to live with Katina's sister in San Antonio. Although the family remained close and George and Maria still saw each other regularly, the separation made losing their mother even more difficult.

"Words cannot describe the feelings I have in me now," George wrote to Maria several years later. "I, too, feel the longing for a mother, someone I can go to, some place I can call home. . . . You are all that I have left. I love you more than any brother could possibly love a sister."[44] George vowed that when he finished college, he would begin working to bring the scattered family together again.

While living with his aunt and uncle, George developed a deeper love of the outdoors. He still visited Galveston frequently, and he still enjoyed fishing its jetties—as he would for the rest of his life. He and Uncle Jimmy also hunted frequently, not just in Galveston but in the woods of the mainland as well. There were more trees there, and having grown up on an island, George was captivated by the oaks, cypress, and other species, as well as the large fig tree that dominated his aunt and uncle's backyard.[45]

Life on the mainland was different. George couldn't simply take off to the beach whenever he wanted, and Lela and Jimmy kept a watchful eye over their young nephew. There were other changes. Uncle Jimmy was Greek, but Aunt Lela was of English and Irish descent, and George had to get used to speaking English at home.[46] Within a few years, though, Jimmy and Lela moved back to Galveston, and George enrolled in Ball High School. After Katina's death, he seemed to apply himself to school with even greater determination, and he had a serious desire to honor his mother's wish that he study medicine at Rice University, an exclusive private school in Houston. "Georgie was real quiet growing up," Galli, his childhood friend, recalled. "He was also the most brilliant boy in high school."[47] He did well at English literature, composition, Latin, history, biology, and science, and he showed a special aptitude for art, although he never cultivated his talent.

In sports, too, George was a loner. Johnny had played baseball with the local American Legion baseball team, the Rattlesnakes, and many of George's friends played football growing up. Galveston may have been an island on the far edge of the state, but it was still Texas, and high school football was a Friday night religion. George, though, didn't have the physical size for the game, and his personality was better suited for a more singular pursuit than a team sport. He gravitated toward tennis, often playing by himself on the island's seaside courts. "You couldn't exactly say he was the runt, but he was quite a bit younger and smaller than his brothers," one of Johnny's childhood friends, Demas Caravageli, recalled. "So, it was real hard for him when he tagged along with us. He just couldn't play sports with the older kids, so George went off on his own to Menard Park and damned if that boy didn't teach himself tennis." The family couldn't afford lessons, so George practiced on his own, showing a quiet discipline and drive to succeed. "You'd see him out there all the time on the tennis courts off the Seawall," Caravageli said. "Sometimes with a friend, sometimes by himself just playing tennis. He got real good at it."[48]

In the difficult years after his mother's death, tennis became George's refuge. He spent hours alone on the court, hitting against a backboard. The routine taught him self-reliance, which not only turned him into a formidable player but enabled him to excel academically as well. He graduated from high school in 1935 with an 87 average, putting him in the top 10 percent of his class, and he was a three-year letterman in tennis.[49] Because he had skipped a couple of grades in elementary school, he finished high school when he was just sixteen years old, though he looked even younger with his round face and shock of thick, black hair.[50]

He was accepted to Rice University and planned to study premed as his mother had wished, but the university told him he was too young to start classes. Unsure what to do next, he took "postgraduate" studies at his high school—a third year of Latin, trigonometry, solid geometry, and advanced algebra. He spent a lot of time on the beach, too, looking up at the stars at night. George loved staring at the heavens, and he had read extensively about the planets and constellations. He wondered what it would be like to travel into space, convinced that it would happen in his lifetime. Despite his mother's wishes, he began to think he might enjoy studying astronomy more than medicine.

Mike, too, honored Katina's wishes to send the children to college, largely with the help of family and friends. "Katina tell me, 'Papa, don't marry; give our children a college education,' and that's what I did," Mike later recalled. "Me, a handsome young man then, stay single; I never marry, but our four children they all have a college education. Me, I never went to school anywhere a day in my life."[51]

As George was getting ready for college, his older brothers were moving into business for themselves. Christie graduated from the

University of Texas (only, he later claimed, because his father sent the dean a case of whiskey), and when World War II started a few years later, he enlisted in the navy and served as a publicity officer.[52] He was known for "hobnobbing with top brass and mingling with celebrities." Those socialite habits didn't change after the war, when he returned to Galveston and opened a beachfront bar and bathhouse called the Beachcomber, which also became the name of his newspaper column. He was known around town for his big, flat-brimmed Panama hat, tilted on his balding head, and the huge cigar that always seemed to be jutting from the side of his mouth. Postwar Galveston's beachfront clubs—the Maceos' Balinese Room in particular—drew headline celebrities like Frank Sinatra, the Marx brothers, and Tony Bennett, and few passed through the island without meeting Christie.[53]

Johnny was equally gregarious and a natural salesman. He would become known for his backslapping demeanor and his greeting of "Hello, hero!" when meeting someone on the street or in a restaurant. (Some claimed that Johnny relied on his trademark salutation because, unlike his father, he was terrible at remembering names.) He graduated from Texas A&M with a degree in chemical engineering and briefly worked for Humble Oil—which would later become part of Exxon—in East Texas. But Johnny had Mike's independent streak, and on a trip back to Galveston in 1936, he met a group of Greek immigrants who had invested in an abandoned well in western Louisiana, which included an old wooden drilling derrick. They offered Johnny a chance to invest in the project for $1,000. They wanted to redrill the old well and try to get some oil out of it. These types of projects were popular with independent oilmen, who could buy old or abandoned wells on the cheap and then try to boost the output by replacing equipment or injecting water into them to force out more oil. The process was risky, but less risky than drilling a new well. At least the well Johnny was buying into had produced oil at one time. But other problems soon arose. "By the time I got to Louisiana, the Greeks had started fighting, because they can't stand prosperity," Johnny recalled. "They started suing each other and I wound up with one lousy well and a lawsuit—and busted."

Undeterred, he formed a new company and named it after his mother—Katina Oil Co. He bought the drilling rights to some land near Vinton, Louisiana, and sold stock in his company to other Greeks back in Galveston—apparently without registering the securities. He built another wooden derrick and struck oil. Then, he drilled a second well that was a dry hole. "I was busted again," he said.[54]

Too young to enter college and with time on his hands, George spent a summer working as a roustabout—basically doing whatever jobs needed to be done on Johnny's rig.[55] Johnny wouldn't let him on the drilling platform because he thought it too dangerous for his kid brother, so George spent six hours a day laying pipe, cleaning up after the rig crews, and doing other odd jobs around the well site. Before he started working for Johnny, George did what he frequently did when confronted with a new challenge: he read all he could about it. He consumed every book he could find on the oil industry and the emerging field of petroleum engineering. "I tried to understand more about the geology," he said. "The field operation intrigued me and the fact that I was good at mathematics had an effect."[56]

The more he studied the oil business the more he was convinced that it offered the best road out of his family's continuing financial struggles. Working at the drilling site was hard, even backbreaking at times, but he loved the thrill of seeing a well come in. He also understood the risks. After George left at the end of the summer of 1936, things again soured for Johnny. None of the wells he drilled over the next few years found oil. His rig crews sued him for unpaid wages, and the local sheriff seized his rig and sold it at auction to settle the debts. Johnny rushed back to Galveston to raise more money from the Greek community and managed to pay his crew and buy back the rig. He got a Greek Orthodox priest to bless the four corners of the derrick before he started drilling, but he still didn't strike oil. For the next four years, he struggled to get by with no car, no money, and a fifteen-dollar suit that seemed perpetually wrinkled despite the family pressing business. He bounced checks "all over Vinton" and ate his meals on credit.[57] In 1939, he spent Christmas alone at his rig, eating a bologna and cheese sandwich because the last restaurant in Vinton had cut off his credit. "Failure, I presume, was written in the books for me," he wrote to his sister, Maria. "Fate, fickle as it is, hasn't smiled spontaneously in these past four years for me. Perhaps someday it may loosen its grip on my downward destiny and let me raise myself from the ranks of a nobody."[58]

Despite his brother's hardships, George had learned a lot working for Johnny. He began to realize that medicine wasn't what he wanted to study. He also may have taken some inspiration from his other brother, Christie, who, in one of his final columns as editor of the student newspaper at the University of Texas, offered some words of wisdom. He advised avoiding journalism because there was no money in it, but he noted that while "there might be a depression for graduates in some fields . . . in the profession of petroleum production, there is a boom."[59]

Even though George had been accepted at Rice and he and Mike had been up to look at the campus, it wasn't the right school for studying geology or petroleum engineering. Instead, he applied and was accepted at Johnny's alma mater, Texas A&M. Aunt Lela drove him to the train station in Houston, and George disembarked an hour or so later in College Station, little more than a bump in the road in the middle of the Central Texas prairie. As he walked across campus, his mind raced with questions it would take years to answer: "What am I headed for? Where am I going? What is my future going to be?"[60]

CHAPTER 4

The Fortunes of Lovesick Freshmen

Texas A&M was an all-male, all-military school. It had a reputation as a rough place, and George Mitchell, a skinny freshman assigned to the B Battery Field Artillery unit, found it intimidating when he arrived in the fall of 1936. A&M was a land-grant university, founded in 1876, with the mission to teach practical skills in agriculture, engineering, science, and fields other than the liberal arts traditionally taught at other colleges. It was also a place where poor kids, many of them sons of farmers and ranchers from Texas's vast rural areas, could get a degree for little money. (To this day, A&M students, known as Aggies, derisively refer to their rivals at the University of Texas as "tea sippers," a reference to the tuition disparity between the two schools and their view of UT as a school for wealthy elites.) George entered A&M two and a half years after Johnny graduated. At the time, few schools offered degrees specifically for the oil business, and few oilmen had been to college. After all, it had been only about three decades since the modern oil industry began on a windswept hillock near Beaumont, about fifty miles northeast of Houston.

The hill was known as Spindletop for the way the wind caught the tree leaves and spun them like a child's toy. In 1901, a wooden derrick punctured the ground, unleashing a stream of oil that shot a hundred feet in the air. For nine days, the Lucas Gusher spewed its fountain of oil, beckoning men and money. Soon both flooded Beaumont, and within a few years, some six hundred companies were doing business there. By 1903, the boom was already playing out, but it didn't matter. Spindletop's success captured the imagination of a new breed of entrepreneurs, modern-day prospectors who swarmed over East Texas just like those who had flocked to California in the middle of the previous century looking for gold. Known for its rolling hills and pine forests, the region didn't resemble Hollywood's barren, tumbleweed-dotted vision of the Texas landscape, but it suffered from drought and sandy soil that made for poor farming. Spindletop, though short-lived, offered hope for the locals and the allure of fast wealth for those who dared to search for oil. In the decade after the discovery, fortune seekers tried to tap into the oil lurking beneath East Texas, but the technology simply didn't allow wells to be drilled deep enough. Geologists at the time became convinced Spindletop was a fluke. There was no oil in East Texas.

Columbus Marion Joiner, a seventy-year-old Alabama native who walked hunched over from the lingering effects of rheumatic fever, didn't believe the scientists. In fact, he decided to drill in East Texas based on the advice of a man who was more soothsayer than geologist. For three years, Joiner managed to scrape together enough cash to keep his rusted, dilapidated equipment running on a small farm owned by a woman named Daisy Bradford. In early September 1930, his well, the Daisy Bradford No. 3, tested positive for oil. There was no gusher, and no one danced in the black rain of liquid gold as men had at Spindletop, but word got out nonetheless and the prospectors flooded back. Almost overnight, a shantytown sprang up and became known as Joinerville. A month later, at 8:00 p.m. on October 3, 1930, the ground surrounding the town began to rumble. The Daisy Bradford No. 3 gurgled and coughed and soon spewed forth a column of oil—a gusher after all. Joiner's gusher made local headlines, but many people thought it was hype or an outright lie, a trick by the promoters who often fleeced investors out of their money with promises of oil wealth. In the next few months, though, more wells came in, and the field, later known as the Black Giant, became the biggest oil discovery in the world. While the rest of the country struggled through the Depression, East Texas boomed.[1]

Joiner—nicknamed "Dad" because his discovery gave birth to the modern oil industry—didn't just make a name for himself; he gave credence to a new breed of businessman. Officially, they were known as "independent operators" because they worked apart from the major oil companies—mostly remnants of John Rockefeller's Standard Oil Trust, which the government had broken up in 1911.

In the new vernacular of the oil business, these independents had another name: wildcatters. They looked for oil where no one thought it existed. They put together deals with less capital—operating on shoestring budgets—and took a smaller slice of the deal than any big oil company would. They had to offer bigger payouts to landowners and investors as a reward for the financing they needed. The hardscrabble nature of the business appealed to a certain type of risk taker, and it offered the prospect of wealth for those who had little means and were willing to hustle.[2] In other words, it was exactly the kind of business that appealed to the sons of Mike Mitchell.

By the time George entered A&M, just six years after Joiner's discovery, the school offered a degree in petroleum engineering. George couldn't decide whether he wanted to study that emerging discipline or the more established field of geology. He decided to do both, working toward an engineering degree but taking as many geology courses as he could.

George's biggest challenge, however, was finding the money to stay in school. Mike's finances had only deteriorated since Katina died, Johnny was still hustling for investors to keep his business going, and Christie, the journalist, "had a poor job" that didn't pay well enough for him to help George.[3] The school may have been cheap, but it charged him $375.25 for tuition, room and board, laundry, and other fees each semester.[4] Almost immediately, George went to work waiting tables in the mess hall, earning twenty-six cents an hour. It wasn't enough to cover his bills, and the administration constantly threatened to kick him out of school.

To supplement his income, George built bookcases that he tried to sell to fellow students. Because of his hardship, he convinced the captain of his battery to allow him to sell candy in the dorm. He couldn't man his makeshift candy store all day because he had classes and military drills, so he arranged twenty or thirty types of candy on a table in the dorm with a sign asking his fellow students to drop their payments in a box. Aggies had a code of honor saying they didn't lie, cheat, or steal, but George's use of the honor system put the code to the test.

He also became the tailor for his unit, using skills he'd learned after school at the Neat Dresser's Club back in Galveston. "Very often, I wouldn't have any money if I didn't make it in candy," George recalled.[5] "They were about to kick me out two or three times." Sometimes football players would take the candy without paying for it or upperclassmen would stiff him for the tailoring he did on their clothes. When his own moneymaking efforts failed, he borrowed from Buddy Bornefeld, a fellow Galvestonian in his unit who came from a wealthy family. "Several times during the year I had to borrow money from Buddy to be able to make it," George said.[6]

In the summers, George returned to work with Johnny in the Louisiana oil fields, which helped him save a little cash. But his second year at A&M was marked by many of the same financial struggles. He was excelling at his studies, but he owed the school twenty-nine dollars a month for his room. He frequently fell behind and after forty-five days, the administration threatened to expel him. When he'd exhausted every other possibility, he would wire his father. "My father never did have any money, but he knew all the people in Galveston," George recalled. "So, he'd take my wire and he'd go over to Sam Maceo and say 'OK, my son is in the top of his class and he needs a hundred dollars.' And so [Sam] would give him a hundred dollars and he'd send me $50 and keep $50."[7]

Maceo may have gotten his money from his family's control of Galveston's seamy underworld, but he was known around town for his generosity. He helped rebuild an African American church that was damaged in a storm and gave "loans" to friends like Mike Mitchell, never seeking repayment. Instead, he would tell people, "someday when you run across somebody who is really in need or is sick, take the money and give him a helping hand."[8] Even if Maceo never told George that directly, the future billionaire would always be grateful for the help he received from Maceo, and he would ultimately repay the mobster's generosity to the benefit of the entire city.

George knew he had to graduate in four years, but the extra geology classes he wanted to take increased the financial pressure. He carried a grueling course load: twenty-three hours a semester. He was in class every weekday until 5:00 p.m. and every Saturday until 1:00 p.m. Even so, he still managed to find time for his favorite pastime, joining the tennis team. A&M didn't offer scholarships for tennis in the 1930s—if it had, George's financial woes might have been over—but he became team captain.

George was well liked in his unit, too, where his fellow cadets called him "Greek." He was still rail thin—160 pounds—and his dark hair was already beginning a retreat that would leave him bald before middle age, but his likable demeanor put him on the fast track for promotion. As a senior, he was named commander of his battalion and held the rank of cadet major in the corps of engineers.[9] He moved into a corner room on the top floor of Dorm 12, a privilege of his rank, and things became a little easier. He was still taking a heavy course load, doing the military drills that were required of all A&M students, and playing tennis in what spare time he had, but his finances improved. The candy business was too vulnerable to losses from theft, and he didn't sell enough bookcases to keep him in school. But George hit on another idea: selling gold-embossed stationery, customized with students' names and the symbol of their unit—crossed cannons for those in the artillery, a horse for members of the cavalry, and so forth. The business took off when George realized the most lucrative market segment was incoming freshmen who missed their high school sweethearts back home.[10]

"By the time I was a senior, I must have been making close to $300 a month selling stationery to those lovesick freshmen," he recalled. "They'd spend their last nickel to write home to their girlfriends." The income was steady enough that George could not only cover his college costs, but even send money home to his father.[11] There would be no more borrowing from the Maceos.[12] He even managed to send a few bucks to his brother Johnny, still struggling in the Louisiana oil fields.[13]

Petroleum engineering was one of the most demanding disciplines at A&M, but George thrived on it and was named a distinguished student. He also loved the more imaginary aspect of geology, which awakened a creativity in him that he had first experienced in art class back in high school. In his junior and senior years, George took several classes from Harold Vance, a petroleum engineering professor who also did consulting work in the industry. Vance's real-world knowledge resonated with George, and Vance's side business gave his best students a chance to gain experience analyzing well logs for Vance's clients. Vance also offered George some advice that would shape his future. "He said, 'if you want to go to work for Exxon (or Humble [Oil] at that time), fine, then you can drive around in a pretty good Chevrolet,'" George recalled. "'But if you really want to drive around in a Cadillac, you'd better go out on your own someday.'"[14]

While George may have been dreaming of Cadillacs, he also gained an appreciation of the risks involved with entrepreneurship. In addition to working with Johnny on his rig in Louisiana during the summers, George began consulting on the engineering of the wells. Johnny was an engineer, too, and he knew the oil business, but he didn't focus on the mechanics of drilling.[15] "George was down there and gave Johnny a few tips on how to dig a well," Christie wrote to their sister, Maria. "It was funny hearing the little shrimp telling Johnny how the cow ate the cabbage. I almost died laughing."[16]

Even with George's help, though, Johnny's finances worsened. He wound up filing for bankruptcy, and because of a piece of property he shared with his father, Mike got dragged into the proceeding as well. Eventually, George would wind up repaying the debt for his family—about $8,000—over five years. He also helped pay for his sister, Maria, to go to college at the University of Mary Hardin-Baylor in the Central Texas town of Belton.[17]

George's senior year, beginning in the fall of 1939, was a heady time for A&M. The school's football team won its only national championship, ending its undefeated season with a 14–13 victory over Tulane in the Sugar Bowl. The team's biggest star was full-back John Kimbrough, known as "Jarrin' John" or the "Haskell Hurricane" after his West Texas hometown. Kimbrough finished second among finalists for a Heisman Trophy and was famous for breaking tackles and creating his own holes in opponents' defensive lines. He was large for a running back of his time—six feet, two inches and 210 pounds.[18] Next to George, he looked like a giant. As captain of the tennis team, George lived in the athletic dorm, and he still had his candy business going. The football players, basking in the fame of their success, devastated the candy store that year. "All these big football players, the national champions, Kimbrough and that bunch, big bruisers, they'd eat that candy and never put any money in," George said.[19]

The campus burst with pride over the championship season that fall, but by the spring of 1940, things were looking more ominous for the graduating seniors. In Europe, the Nazi blitzkrieg rolled through Czechoslovakia and Poland, and Hitler set his sights on Western Europe. In a military school like A&M, graduating seniors faced the growing likelihood of combat. "The war was getting close, so the Army was recruiting some of the top students to become permanent officers at that time," George recalled. "But I didn't want that. I knew the war was imminent, maybe in a year or two. I didn't want an Army career. I wanted to get out in the oil business and get going."

He graduated at the top of his class, with a grade point ratio of 2.37 on a three-point scale. In all, he accumulated 390 grade points during his four years in college even though he needed only 160 points to graduate.[20] The Petroleum Engineering Club gave him a gold watch, a traditional gift for the most outstanding student each year, and he was a letterman in tennis three years in a row.[21]

As graduation neared, Mike and Christie sent him a congratulatory telegram: "She is proud of you today."[22] Indeed, George had accomplished the goal his mother had set for him—graduating from college. Now, he was ready to pursue his own dreams.

While many young men at the time were eager to enlist, George turned down the army's offer of an officer's commission and began interviewing for jobs with oil companies. The country was still shaking off the Depression, and jobs were hard to find despite George's academic standing. He had some opportunities to work for oil companies overseas, but he wanted to stay closer to home.[23] The only deal that would keep him near Galveston came from Amoco, one of the country's biggest oil companies, which had split off from Rockefeller's Standard Oil. Amoco put some conditions on the hiring. George had to report immediately to a well in southern Louisiana where the company had just had a blowout.

He barely had time to pick up his diploma, which came with a reserve commission in the army. He missed the final review, when lowerclassmen assembled on the drill field and submitted themselves for inspection by the graduating seniors. For members of A&M's tradition-rich Corps of Cadets, the event was almost as important as graduation itself—if not more so—and George had eight hundred students under his command. His father, aunts and uncles, and brothers and sister had all come for the ceremony, but Amoco insisted he report for work the next day if he wanted the job. He skipped the review and Johnny drove him to Lafayette. It wasn't the way he wanted to begin his career in the oil business. "It made me so damned mad, but I had to do it because jobs were very scarce, and it was a good opportunity," he said.[24]

The opportunity may have been good, but the job itself was hard and the pay was $126 a month—less than half what he had made selling stationery in the dorm at A&M. To make matters worse, his boss, a man known as "Wooly Red" Bedford, seemed to enjoy making life miserable for his newest employee. It was Bedford who had insisted the company needed George on-site immediately, but when George arrived, he found he had nothing to do. Bedford didn't need him; he simply "wanted to show me that you had to be right on the ball."

The first winter, George was assigned to work for three months as a roughneck on a rig in Hackberry. Roughnecks handle the hard manual labor on rigs, such as moving drill pipe into place on the derrick. It was the sort of dangerous work that Johnny had forbidden him to do in high school. That winter was unusually cold for Louisiana, and he had to work through rain and sleet. "It was dangerous as hell," he recalled. "I thought, 'my God, if I thought I'd have to do this all my life, I'd go jump off this damned rig right now.'" After several months of roughnecking, though, he began overseeing the production on Amoco's twenty-two wells in southern Louisiana. Finally, he was using the engineering skills he'd learned in college. He was living with a Cajun family in Jennings and working out of the office in Lafayette about forty miles away. While Bedford may have been a difficult boss, George's immediate supervisors treated him well, and he liked working with the other engineers. He was gaining hands-on field experience that would be critical if he planned to follow Vance's advice and strike out on his own someday.

The war in Europe continued to rage, a constant drumbeat in the background as George tried to establish his career. In June 1941, the army ordered Johnny to report for active duty, and George knew his time was coming. After all, they had both received officer training at

A&M and were now in the reserves. The United States wasn't at war yet, and Johnny didn't want to be a soldier any more than George did. The difference, though, was that while George had a steady job with Amoco, Johnny was still trying to make it on his own. "I wasn't faring too well, either," Johnny would later recall. "Owning one small drilling rig and a few barrels of oil, I was barely surviving following a string of dry holes."

Despite his years of bad luck, Johnny had kept drilling around Vinton, stalling the army on its request that he report. On December 7, as he pulled up to the derrick on his only oil well, the news flashed on the car radio that the Japanese had attacked Pearl Harbor. Stunned, Johnny walked up the derrick steps only to find when he got to the rig floor that the well was yet another dry hole. He ordered the crew to stack up the drill pipe for storage and left his rig behind to report for duty.[25]

By then, George had reported, too. His orders came through three months after Johnny's, in September, and he left a month later.[26] His desire to "get going" in the oil business would have to wait.

CHAPTER 5

"Call Me Mitch"

Despite George's training in A&M's Corps of Cadets, the army wasn't interested in his artillery skills. It wanted his engineering talent. He was just twenty-two years old when he entered the service as a second lieutenant, a position that could have cost him his life if he had been assigned to an artillery unit and sent into combat. Instead, he joined the Corps of Engineers and went to work at the San Jacinto Ordnance Depot in Houston. The Corps was responsible for building the 5,000-acre complex, which included munitions storage and a ship channel. Most of the work was done by private companies, and George was one of about a half dozen army officers overseeing the project, which included managing some two thousand civilian workers. The role gave him his first taste of running a large operation.

In late November 1941, less than two weeks before the Japanese attack on Pearl Harbor and before he had really settled into his new job, George got a weekend pass. He took the train to College Station for the football game between Texas A&M and its archrival, the University of Texas.[1] A&M's national championship was two years in the past, but around campus, winning the traditional end-of-season game against the Longhorns mattered more than post-season play. A&M was having another great season. The team was ranked number 2 in the country and had already clinched the Southwest Conference. The Longhorns hadn't won in College Station in eighteen years. A fortune-teller in Austin suggested that the Longhorns could break the jinx by burning red candles before the game. Perhaps that was the reason that the Longhorns upset the Aggies 23–0. Heading back to Houston after the game, George, still wearing his military uniform, settled into his seat and surveyed the other passengers.[2] He noticed two couples sitting nearby. One of the women seemed annoyed with her date, who had clearly had too much to drink. He assumed the two women were related because they looked identical, with dark, shoulder-length hair in the same style, and an easy smile.

Cynthia Woods was, indeed, not enjoying the train ride. Her twin sister, Pamela, had set her up on a double date, and the boy had been drinking whiskey from a flask most of the afternoon. She turned to Pamela's date and whispered quietly, "Can't you find a poker game to lose this guy in?" The two men headed for the card car, and George spotted his chance.[3] He wasn't particularly tall, but he cut a dashing figure in his uniform. He had dark, intense eyes, and the roundness in his cheeks had disappeared as he entered his twenties. George may have been naturally shy, but he knew opportunity when he saw it. He walked over to the two women and introduced himself. "The first thing he said was 'I'm George Mitchell—call me Mitch.' I thought he was cute," Cynthia recalled.[4] Perhaps he made up the name on the spot to impress the women. Most of his friends at school called him "Greek," and back home in Galveston he was still "Georgie." In later years, people knew him as "George" or "Mr. Mitchell." But to Cynthia and her sister, he would always be Mitch. As the train arrived at Union Station in downtown Houston, George, still sitting with the two women, very deliberately checked the arrival time by flashing the gold engraved pocket watch he'd received as the top engineering student at A&M the previous year. By then, he'd already gotten both of their phone numbers.[5] That night, Cynthia would write in her diary: "Met a cute soldier today."[6] George went on a date with Pamela first, but eventually, he felt drawn to Cynthia. "You couldn't tell them apart," he said. "But I noticed when I danced with both of them, with Cynthia, something about her was a little different."[7]

Cynthia loved art and design, and long before segregation ended she was a supporter of equality. Her distaste for the institutionalized racism of the 1940s left her out of step with the prevailing social norms of the South. She was born in New York on September 24, 1922. Her father, Clinton, worked in advertising, and his marriage to Cynthia's mother, Loretta, was not a happy one. Clinton was the son of a barrel maker and carpenter in western Michigan, and Loretta was one of eight children from a family of Irish Catholic farmers in central Illinois. They'd married in a Catholic church in the Chicago area in 1909 and moved to New York. By the time Loretta gave birth to twin daughters, the couple was living in the Bronx.[8] Cynthia remembered the apartment as bleak, and she would often move lights around or change decorations to make it prettier. Perhaps it was her artistic nature or perhaps she was trying to push away the unhappiness that surrounded her.[9] Her parents fought frequently, and often bitterly.

When the girls were about seven years old, Loretta told them their father had died unexpectedly. There was no funeral, no visits from family or friends, no flowers or other expressions of sympathy. Instead, Loretta bought Cynthia and Pamela black mourning dresses, and they boarded a train and moved back to Loretta's hometown of Carrollton, Illinois.[10] About the time the girls entered high school, they moved again, to Jacksonville, thirty-five miles away.[11] The town was midway between the Mississippi River and Springfield.

As twins, Cynthia and Pamela frequently wore the same clothes and hairstyles.[12] They loved growing up in the progressive small town, which was home to two academic institutions—Illinois College and MacMurray College, both founded in the 1800s. Illinois College's first president, Edward Beecher, was the brother of author Harriet Beecher Stowe and an outspoken abolitionist. The college became an engine for the abolitionist movement, and during the mid-1800s Jacksonville sheltered hundreds of escaped slaves fleeing the South on the Underground Railroad.[13] The town's reputation for education resulted in the nickname "Athens of the West," and by the late 1930s, it was common for young women in Jacksonville to attend college.[14] The town's history was still obvious by the time Cynthia and Pamela arrived. They attended Newton Bateman Memorial High School, which even in the late 1930s was integrated. Cynthia loved living in Jacksonville, and her high school years undoubtedly helped shape her views on everything from education to racial equality.

Cynthia was editor in chief of the school yearbook, the *Crimson J*, and she won praise from fellow students and the faculty. "Your work has been outstanding while in high school," teacher Helen

Holland wrote in her senior yearbook. She and Pamela excelled at their studies and were popular. The Carrollton and Jacksonville high schools played each other in basketball, and Cynthia and Pamela realized at one point that between them they had been asked out by every boy on both teams.[15] By her senior year, Cynthia was dating the kicker on the football team, Basil Baumann Sorrels, known as "Bas," who took her to the prom.[16] Then, just before they graduated, Loretta told the girls they were moving back to New York. After graduation in 1940, they once again packed their things and boarded a train, but instead of arriving in Manhattan, they wound up in Houston. It would be almost forty years before Cynthia discovered the reason for the sudden change of plans and the truth about her father.[17]

In Houston, the girls got jobs as stenographers to support their mother. They also enrolled in night classes at the University of Houston, where Cynthia studied art and planned to go into advertising like her father.[18] Known as a "vivacious lady," she was active in the glee club and helped put on plays, showing an interest in the performing arts that would persist for the rest of her life.[19] In 1943, she was listed in *Who's Who Among Students*.

As Cynthia worked and pursued her art degree, she and George dated for about two years. The war hung heavily over their courtship. George worried about an overseas posting, and he didn't want Cynthia to wind up a war widow. "Casualty rates were so high during wartime, I thought no one should get married," George later recalled.[20] Cynthia disagreed, but George continued to wrestle with the uncertainty. "Everything is in such chaos, I know not what course to follow," he wrote to Uncle Jimmy in late 1942, after he and Cynthia had been dating about a year. George, of course, was removed from the worst of the chaos, but he was getting reports from his brother Johnny, who was a field artillery officer in Europe with Patton's Third Army.[21]

George, however, stayed close to home. After he finished the job at the ordnance depot, his superiors assigned him and most of his team to the Dickson Gun Plant, built in 1942 on the banks of Houston's Buffalo Bayou. The plant, which had about 1,350 employees, was operated by the Hughes Tool Company, an oil-field services outfit founded by the father of eccentric billionaire Howard

Hughes. George oversaw the production of various caliber gun tubes, working ten to twelve hours a day with little time off.[22] He liked the challenge of expediting the manufacture of guns for the war effort and didn't complain about the long hours. As he saw it, the faster they could get guns to the Allied troops, the faster the war would end.[23]

When that assignment concluded, the prospect of an overseas posting loomed once again. Several colonels attempted to put him on projects such as rebuilding the Burma Road to China after it was recaptured from the Japanese, but his commanding officer was so impressed with George's work that he refused to let him go. It didn't hurt that his superior also liked playing tennis almost as much as George did, and he found George to be a worthy opponent.[24] Tennis, as much as his engineering skills, may have saved George from combat.

Instead of Burma, George was sent to the Galveston office of the Corps—Cynthia joked that by crossing the bay he'd finally gotten an overseas posting—and he spent the remainder of his army career in military production, overseeing work on buildings at Austin's Bergstrom Airfield and the New Orleans Army Airfield on the banks of Lake Pontchartrain. He helped with the construction of an Alaskan pipeline that would pump more crude from a field developed by Imperial Oil (which later became part of what's now ExxonMobil) to a refinery that turned it into fuel for the war effort. He also ran a project to build prefabricated military hospitals for shipment overseas, and another working with oil-field services company Schlumberger to develop a sonic test for finding nonmetallic mines.[25] By then, he had a regular staff of about forty reporting to him.[26]

By 1943, it was clear that George was likely to spend the war stateside, and the course he should follow became clearer. He asked Cynthia to marry him. Cynthia's twin sister, Pamela, got engaged about the same time to Raymond Loomis, a lieutenant in the Army Air Corps. The couples had a dual ceremony on Halloween in Alexandria, Louisiana, where Loomis was stationed.[27] Both officers wore their dress uniforms and the women wore matching outfits of knee-length skirts and jackets. George was twenty-four and Cynthia was twenty-one. George had been promoted to first lieutenant by then, but his pay wasn't much better than it had been in Louisiana. He was making $167 a month.[28]

The Japanese surrendered in August 1945, but George couldn't leave the army. Promoted to major, he still had responsibility for the engineering projects he'd overseen during the war. Closing the contracts and receiving his discharge took until May 1946.[29]

In the meantime, Johnny had returned from Europe. "All I could think about was my little rig," he said. Despite his struggles before the war, he was determined to make it in the oil business. He borrowed a car and drove to Vinton, but he found his lone rig rusted and crumbling. "I had to be the sickest man you ever saw," he said. "I'd dreamed about it for almost 40 months, and I'd dreamed about the wells I was gonna drill, the things I'd missed."[30]

George was still finishing his work for the army, but once again he began consulting with Johnny, and he and Cynthia had settled into a small apartment in Galveston. Housing had been in short supply in the final years of the war, and the young couple had rented a tiny one-bedroom from the family of Johnny's childhood friend Demas Caravageli. Caravageli warned them that the apartment was too small for children. "That's okay because we don't plan to have kids anyway," George replied. By the time he was discharged, George and Cynthia already had one daughter, and another was on the way. Their stay in the apartment wouldn't last long.

On the island, Cynthia found herself thrust into Galveston's Greek culture and inundated with George's extended family—his aunts and uncles; his brothers, Johnny and Christie, both back from the war; and of course "Papa Mike," who was still prowling the poker tables and chasing women. When Cynthia gave birth to their first daughter, Mike wanted her named after her maternal grandmother, in keeping with Greek tradition. (Katina's mother's name was Marigo.) Cynthia refused, naming the child after her twin sister, Pamela, instead.[31] Cynthia and Mike were opposites in many ways—she was well educated and appreciated art, and she had distinct ideas about proper character and behavior; he had no schooling and had survived by his wits, gambling and engaging in questionable business deals to find money where he could. While she avoided being alone with Mike, she still adored him. "You'd think my grandfather would have just made her skin crawl, but she thought he was funny and lively," said George's daughter Sheridan Lorenz. "In the summer, every single day, we would pick him up and go get ice cream."

George was forgiving of Mike's transgressions. Late in life, Mike would say that he regretted not treating Katina better, but George, despite his devotion to his mother, looked past his father's flaws. To him, Mike was a survivor. He had risked everything to come to America, and he had found a way to provide for his family and send four children to college at a time when many immigrants didn't have that chance. It might not have been pretty, but he'd done it. Now, George was thinking about taking risks of his own.

He could see the pent-up demand in the Oil Patch. US companies had done little drilling during the war because the steel needed for drill pipe had been diverted to the war effort. The country was ready to get back to work, and oil companies were primed to drill again. Across the country, people began moving to the suburbs as the post-war housing explosion encouraged a burgeoning love affair with the automobile. The new American ideals of suburban living and a car in every driveway both relied on one thing: oil.

Some of the people he'd worked with on the pipeline in Alaska had formed a company and wanted George to help manage it. He turned them down. He longed to revisit his first love—engineering and geology. Amoco wanted him back, too, but he felt he had lost a lot of time during the four years he was in the army. Many of the people who'd started at the same time he had, and hadn't gone to war, had been promoted. He didn't like the idea of starting at the bottom of the corporate ladder.[32]

George remembered what his professor Harold Vance had told him about Chevies and Cadillacs. He didn't need a Cadillac, but he did need money to support his growing family. He had only one year's work experience, but he decided to go into business for himself. If he didn't make it, there was always a corporate job at Amoco.[33] "I just said to myself, 'The hell with it, I'm going to be a consultant.'"[34]

CHAPTER 6

"You're Going to End Up a Bum"

George was convinced he could make it on his own in the oil business, displaying a youthful confidence that he would later dismiss as naïveté. He managed to convince a half dozen oil-men around Houston to pay him fifty dollars a month as a geology and engineering consultant. The war had taught him some important lessons. From his vantage point behind the lines, he got to see the inner workings of the American war effort—not the combat operations but the logistics that made everything else work. He realized that despite the celebration of battlefield heroics, victory was rooted in these behind-the-scenes services that ensured that goods and materiel got where it needed to go. The war taught him to think big and appreciate the details.[1] At the time, the oil business was still oriented around each individual drilling project. The business, though, was maturing, as Johnny was discovering in Louisiana.

"I came back and found a different world," Johnny said. "It was a new ball game in '46. You have to raise money, you got to drill wells with capital." And capital was in short supply. After finding his rig rusted and useless, Johnny returned to Galveston and sold beer, cold drinks, and hot dogs on the beach. He drove around town in a maroon 1946 Ford and was known for wearing short khaki pants—a style he'd picked up from Australian officers during the war and that he thought would be more practical for Houston's hot, humid summers.[2] The eldest Mitchell brother, Christie, had settled into a similar lifestyle, running a beachfront bar, sponsoring frequent "bathing beauty" swimsuit contests, and writing "The Beachcomber," his gossipy column about the social goings-on across the island.

Despite his prewar travails in Louisiana, Johnny missed the oil business terribly during the war.[3] He was thirty-one years old and dead broke. Like his father, Johnny could find money, but he couldn't keep it. He was a colorful character, and people were drawn to him. "Uncle Johnny was part human and part tornado," recalled George's daughter Sheridan Lorenz. "He was . . . one of the most buoyant, generous, fun-loving and original people I have ever known."[4] Johnny's unit had seen some of the worst fighting in Europe and had endured some of the highest casualties, and the soldiers under his command praised him for keeping them safe. Johnny had little tolerance for military protocol and was frequently insubordinate in standing up for his troops. He also managed to find time to entertain young women from Ireland to Germany and documented his adventures in a book, *The Secret War of Captain Johnny Mitchell*. "He endured some of the most hellacious battles in the European theater and came home and wrote a book about his girlfriends," said his nephew Kirk. He was, after all, Mike Mitchell's son, and the one who was most like his father in both appearance and habits.

When he wasn't hawking refreshments to tourists, Johnny was plotting his return to the oil fields. He needed a lot more money to get back in than he'd needed to get started before the war, but fortunately, Galveston's most accessible source of capital was also his old barber, Sam Maceo. By then, the Maceos' gambling empire not only covered all of Galveston but stretched across the bay and onto the mainland. The Balinese Room had gained a national reputation and had become the center of nightlife on the island. The Maceos still had law enforcement in their pocket—although the feds had tried unsuccessfully to shut them down—and they gave so generously to every church in the county, both Catholic and Protestant, that after Sam's death in 1957, citizens would lobby against closing the Balinese. (The Balinese, the subject of the 1970s song of the same name by the Texas blues band ZZ Top, was added to the National Register of Historic Places in 1997. It was destroyed by Hurricane Ike in 2008.) When state authorities finally shuttered the Maceos' operation, they found it was generating $5 million a year.

But in 1946, Sam Maceo wanted to become an oilman, the new

status for wealth in Texas. He'd given Johnny some money before the war to help get him started, but that was just loans for which he never expected repayment. It was similar to the money Maceo had lent Mike to keep George at A&M. Now, Maceo was ready to become a serious investor. He could see oil was the future, and he was ready to make a second fortune in a legitimate business. Slowly, with Maceo's help, Johnny acquired some mineral rights and began drilling again. And while Maceo knew nothing about how to drill a well, he was more than a passive financial backer. He had a station wagon outfitted with a stove in the back, and he would drive to the well site and cook spaghetti for Johnny and his crew. "He was the nicest partner I ever had," Johnny said.[5]

Johnny's capital shortage was easing, but he still couldn't overcome the biggest impediment to success in the oil business: he couldn't find oil. As much as he loved the thrill of drilling, every dollar he raised from Maceo and others wound up paying for dry holes.

George, on the other hand, had an almost intuitive sense that, combined with his geology training, gave him an uncanny ability to spot oil in the ground. It was as if he could imagine the treasure pooled in reservoirs deep below the surface and divine the geologic anomalies that could point it out. He drew his consulting clients from a network of family friends and people he and his brothers knew from Galveston, and many of them were also investing with Johnny. One of George's earliest clients was Floyd Karston, a fair-sized operator in the Houston area who had three backers—Morris Rauch and Louis and Harry Pulaski. Rauch and the Pulaskis didn't trust Karston's expertise, so they hired George to examine the engineering and geology on the deals. Karston, however, was a better operator than his investors gave him credit for, and with George's help, he drilled a lot of successful wells.

Rauch ran a shop that provided oil-field supplies to small operators. Johnny was a frequent customer because Rauch would extend credit to oilmen who didn't have money. "He helped me like a father," Johnny said. "I'd walk in there, and I'd say 'Mr. Rauch, I need a hundred dollars to get back to Louisiana,' or 'I want a hundred foot of surface casing.'" As a Jew in the oil-field supply business, Rauch, too, was taking a gamble. Many larger companies and operators wouldn't do business with him, and by extending credit to small-time wildcatters like Johnny, he built a loyal, if risky, customer base. Slowly, Johnny began to assemble a syndicate of investors, and through Rauch, he made other connections in Houston's Jewish community that would prove beneficial a few years later.

One of George's most prominent consulting clients was the legendary Houston oilman Glenn McCarthy, on whom the Jett Rink character in *Giant* was based. McCarthy was George's opposite in almost every way. A bourbon-swilling wildcatter prone to fistfights, he'd been kicked out of Texas A&M for hazing. He had dark eyes and an Errol Flynn mustache, and he liked to throw his money around and hobnob with Hollywood starlets as his career swung from fortune to near bankruptcy and back. McCarthy developed a reputation as a great oil finder, plucking black gold from the swamps and marshes around Houston, and by the end of the war, he'd amassed a fortune of $50 million. By the time George went to work for him, McCarthy was more focused on other interests—owning a 15,000-acre West Texas ranch, flying planes with his friend Howard Hughes, and pouring much of his wealth into the ostentatious Shamrock Hotel in Houston, which included the world's largest pool with, McCarthy claimed, the world's largest towels. The Shamrock brought Hollywood to Houston, and it would eventually cost McCarthy his fortune, but George never got caught up in the glitz and glamor of his client. Instead, he quietly evaluated deals for McCarthy, whose wealth continued to grow during the time George worked for him, topping $70 million.[6]

Meanwhile, Johnny had gotten in touch with H. Merlyn Christie, whom he had known since high school. Merlyn had frequently visited Uncle Jimmy's bathhouse, where Johnny had worked as a lifeguard. Merlyn was taller than Johnny, with deep-set eyes and dark black hair that he wore slicked down and sharply parted. In the years since high school, he had become a successful stockbroker with a wealthy client list, and like Sam Maceo, he felt the call of oil.[7] He wanted to move away from brokering deals and involve himself in drilling wells. He'd formed a company with two other partners, Roxie Wright and Jim Gray, and they were looking for places to start. Wright was a drilling contractor who owned his own rig, which he contributed to the company in exchange for getting to name the firm Roxoil. Merlyn asked Johnny to join them and then made another suggestion: Why not have your brother George evaluate our drilling prospects?

Roxoil soon changed its name to the less imaginative Oil Drilling Co., and after three or four months, Gray decided the venture had little promise and moved to Oklahoma. George agreed to buy Gray's stake for $9,000, but because he didn't have the money, he borrowed it from the bank. His credit wasn't good enough to get the loan on his own, so his consulting clients, Rauch and the Pulaski brothers, endorsed the note. Merlyn owned half the company, George owned 37.5 percent, and Johnny owned the remaining 12.5 percent.[8]

The company still had only one rig and almost no income, but Merlyn Christie was determined to do oil deals. He relied on George to handle the geology and put Johnny to work peddling the projects to potential investors.[9] It was a perfect fit. The personalities of George and Johnny were as different as those of their mother and father, but Merlyn Christie had found a way to combine their strengths. George would spend hours on his own reviewing maps and geological data, while Johnny wined and dined wealthy investors.

Cynthia didn't like the idea of George joining Johnny in business. She thought her brother-in-law's wild ways would sink the company before it got started. "If you're going in with your brother, you're going to end up a bum," she told him. Perhaps because of Cynthia's cautions, George maintained his own consulting business on the side, which brought in about $300 a month.[10] Cynthia was pregnant again, the growing family needed more space, and Oil Drilling would be run from an office in downtown Houston. It was time to move inland. George's aunt, Eleftheria Eleftheriou, known as Lefty, owned some apartments in Houston, and the couple moved out of Demas Caravageli's tiny island flat into a two-bedroom in central Houston. Over the next few years, Cynthia would give birth to three more children—two sons and a daughter—and her mother, Loretta, who was suffering from cancer, would also move in with them.[11] The new apartment would get small in a hurry.

Cynthia had maintained a close relationship with her mother, and by the time Loretta came to live with the Mitchells, she was delusional from her illness. She insisted that Cynthia examine the contents of a mysterious trunk. Cynthia didn't pay much attention to what her mother was saying and never got around to opening the trunk before it was destroyed in a flood years later. It would be almost three more decades before she understood what her mother was trying to tell her.

Cynthia's father hadn't died in New York when she was child. Loretta had left Clinton—as Catholics, they would have been unable to get a divorce—and had taken the girls with her. The sudden move to Houston from Illinois when Cynthia and her sister graduated high school had been a desperate attempt to keep Clinton from seeing his children. Clinton, who by all accounts was mild mannered, wrote to Loretta constantly, begging to see the girls. She never responded, but she saved all the letters in the trunk. In 1959, Clinton edited a book, *Ideas That Became Big Business,* and he died in Maryland in 1965.[12] By the time Cynthia learned what had happened, both her mother and father were long dead. "To find out that her mother had kept this secret about her father—that was really damaging to my mother," Sheridan Lorenz said. "It was too late to get to know him, and it was too late to re-evaluate her mother."[13]

◈

George's decision to buy into Oil Drilling Co. represented one of the fundamental paradoxes of the oil business at the time. Energy companies were typically run by businessmen who were politically conservative yet would take huge financial risks. Oil Drilling was run by a failed wildcatter, a geologist with one year's experience, and a small-time broker who had traded a few oil deals. But the combination worked. George's oil-finding skills overcame Johnny's bad fortune in wildcatting, and Johnny peddled George's successes to investors, who gradually became more impressed with the company's growing track record.

They set up shop in a small office in the Esperson Building, one of the first skyscrapers in downtown Houston. Every morning, anyone who was anyone in the local oil business would help fill the one hundred seats in the building's ground-floor drugstore, drinking coffee and cutting deals. George didn't go downstairs; Johnny did. Among the brokers and industry players, Johnny was in his element, working the room, slapping backs, and convincing people that Oil Drilling had deals worth investing in. "He was the glad-hand man," George recalled later. "Then, if they liked it, they'd come up and I'd show them the geology."

It wasn't as easy as the brothers made it sound. To understand the geology of the deals he and his brother were pitching, George needed to study well logs—records made by drilling into the earth and recording the geologic properties of the rocks. A monitor, or logging tool, was lowered into the hole on a wire. It sent out electrical signals that created a chart, similar to an electrocardiogram that doctors use to check heart activity. Studying the readout—hairline, staccato undulations that could stretch for dozens of feet—would give George an idea of the best locations and depths for striking oil.

To truly understand the contours of the geology in a particular region, George needed about twenty logs for each project, and gathering enough logs would cost hundreds of dollars. Oil Drilling didn't have the budget for that kind of research. Instead, George turned to a friend for help. John Todd ran Cambe Blueprint, a small library for geological data on Milam Street in downtown Houston, a short walk from Oil Drilling's offices. Cambe bought logs from oil companies, reproduced them, and sold them for ten dollars. George convinced Todd to let him pick up the logs at closing time and keep them overnight. "I would borrow the electrical log, so I could do the geology at night and work until two or three in the morning," George said. "Then I'd take them back the next morning, [so] I wouldn't have to pay for them."[14] The trick gave George access to the data he needed to make a convincing case for the geology.[15]

The brothers drilled their first well in 1946 at High Island, on the north side of Galveston Bay (despite its name, it's a small community on the Bolivar Peninsula), and struck oil.[16] Typically, Oil Drilling would work on farm-out deals—small pieces of acreage on which a major oil company would offer to let them drill a well. Because the big company was already drilling in the area, it increased the chances of striking oil, but George still had to pinpoint the best sites by studying the logs and other geological data. The big oil company would offer them a parcel of land and pay a percentage of the drilling costs. Oil Drilling would find its own investors to cover the rest of the expenses. For example, a company like Gulf Oil might offer three hundred acres and put up $10,000 for drilling a well. But a typical well cost $30,000 to drill, so George and Johnny would find investors to contribute the remaining $20,000. Investors got paid a percentage of the proceeds from any oil that the well produced based on the amount of their investment. Gulf would hold what's known as an overriding royalty interest—meaning it got a percentage of any production off the top. Gulf essentially paid one-third of the operating costs of the well, and Oil Drilling paid two-thirds, using money raised from investors. If the well was successful, Gulf kept one-third of the proceeds and Oil Drilling distributed the remaining two-thirds among its backers, keeping a small piece for itself. The strategy enabled companies like Gulf to develop their fields more quickly without having to shoulder all the drilling costs themselves. But it didn't leave much for companies like Oil Drilling. Most wildcatters tried to hang on to at least a one-eighth interest in the deals they put together. As it fought to establish itself, Oil Drilling settled for one-sixteenth. If a well succeeded, Oil Drilling would get its fee. If it failed—if it was a dry hole—everybody lost.[17]

When a well came in, George and Johnny raced to the bank. Once a well was producing, banks would lend against it, so George and Johnny would borrow $8,000 to $10,000 against their one-sixteenth interest to cover the costs of completing it—capping the well and putting in the equipment needed to capture the oil for sale. Larger companies could arrange loans against a package of such deals, but in those days, George and Johnny were living on the oil-field equivalent of hand to mouth. It was yet another paradox of the oil business—George and Johnny needed the money to prove they could be successful, but the money came at such a high price that financial success came far more slowly than their drilling success. Still, the strategy worked well enough that in a few years, George was making about $2,000 a month in royalties—enough to move

Cynthia and their children out of the two-bedroom apartment and into a four-bedroom house on North MacGregor Way, just south of downtown Houston.[18]

Cynthia had longed for a house, and she loved the residential area known as Riverside Terrace near the University of Houston. When she had first moved to Houston, she had passed by the neighborhood every night on her way to classes, and she always dreamed of living in a place like that. Developed in the 1920s, it was still considered a premier neighborhood for Houston's more affluent residents thirty years later. George and Cynthia would have four more children while they lived there—all boys—and the backyard included a log fort and a concrete racetrack where the children could ride their bikes. By the time the family had settled in, George was enjoying more success than anyone in his family had known until that time.

As the number of successful deals rose, George and Johnny increased their percentages, first to an eighth and then to a fourth, capitalizing on their perfect combination of intuition and fundraising. The more deals they did, the more George demonstrated an innate talent for detecting oil, often when others said it wasn't there. Johnny was a natural-born salesman and convinced investors that if they gave Oil Drilling money, George would find oil.

Johnny built his investor syndicate around a core group that he dubbed "The Big Nine." They included members of some of Houston's most prominent Jewish families—Galveston attorney W. N. Zinn; the Weingartens, who ran a local grocery chain; the Oshmans, who operated sporting goods stores; and of course, Rauch and the Pulaskis. More than anyone, they were critical in establishing the Mitchells in the oil business, and they provided a reliable source of funding for Oil Drilling. Oil Drilling, in response, provided growing returns.

While Johnny was the master salesman, George attracted investors by impressing them with his geological prowess. In 1949, legendary Houston oilman R. E. "Bob" Smith was looking at a deal in the Texas Panhandle. Smith was a larger-than-life figure both in the oil business and in Houston. Born in the East Texas town of Greenville, he began working for Gulf Oil in the booming

East Texas field before he graduated from high school. He spent a decade playing semiprofessional baseball. When he wasn't on the ball field he was working in the oil field for a host of companies, including Gulf, the Texas Company (later Texaco), and Humble (later Exxon). "I figured I got fired from at least five jobs besides the ones I quit," he once said. "The trouble was I couldn't take orders unless they were properly given." After serving in the army in World War I, Smith wound up a district manager selling oil-field equipment in Oklahoma before getting fired once again, at age twenty-seven. Then, opportunity blew in with a bitter north wind that whipped across the Oklahoma plains in the mid-1920s. A discouraged driller from Texas was shivering in the frigid blast, complaining about the weather and longing for warmer climes. "If I could sell my two rigs today, I would do it, just so I could get out of this cold country and go home," he said, within earshot of Smith. Smith approached him with an offer. "I'd like to buy your rigs if the price is right," he said. The two agreed on $25,000 for both rigs, even though Smith didn't have any money. The next morning, he went to a local banker and convinced him to finance the purchase. Smith used the rigs to drill a couple of wells, both of which were successful. He paid off the loan and walked away with $4,500 in his pocket and a new outlook on his career. He would never be fired again. From then on, he would work for himself.

Smith, who lifted weights regularly well into his seventies, built his fortune in oil largely on intuition.[19] Unlike George, he never went to college or studied geology, but he could find oil by looking at surface changes in the land and, some said, by using pure instinct. By the time he was looking at the Panhandle deal, Smith was already one of Houston's wealthiest residents and biggest landowners. In 1957, *Fortune* magazine estimated his wealth at about $200 million, and he would later cement his legacy by helping to bring to town the team that became the Houston Astros and leading the effort to build the world's first and most famous indoor stadium, the Astrodome.[20]

For all his success in finding oil, though, the Panhandle deal puzzled Smith. Never one to dwell on details, he decided to share the particulars among some young geologists in Houston to see what they thought. Smith was one of Merlyn Christie's brokerage clients, and Roxie Wright had done some drilling work for him, so he shared the Panhandle prospect with Oil Drilling. Most of the geologists who looked at the play told Smith to walk away, but George told him to drill. Struck by George's certainty, Smith took his advice and wound up making the biggest discovery of his career. Of the first eighty-six wells he drilled in the field, all but two struck oil.

In the ensuing two decades, Smith held a 25 percent stake in almost every project George and his company did. Having made one fortune finding oil on his own in East Texas, Smith would make another simply by investing with George. And while Johnny and Merlyn Christie handled most of the investors, Smith would deal only with George.[21] In turn, he taught George something about managing success. He had made a lot of money in oil, but he had diversified his holdings, becoming a major investor in real estate. "He used to say his first fortune came from inside the land and the second was built on it," George recalled. It was a lesson George would take to heart as his own wealth grew. George's fortune, though, wouldn't come from Bob Smith. It would come from a Chicago bookie, and it would set in motion a chain of events that five decades later would change the landscape of the energy industry in the twenty-first century.

CHAPTER 7

The Bookie and the Big Bet

George was sitting at his desk in Oil Drilling's cramped office in the Esperson Building when the phone rang. On the other end was "General" Louis Pulaski. Pulaski had never been in the military, but people called him "General" because, the story went, he had been named for a famous Polish military officer who served the Americans in the Revolutionary War.[1] "I've got a deal for you," Pulaski said. "My Chicago bookie has a deal in North Texas, and he wants you to look at it."[2]

Pulaski and his brother, Harry, were among Johnny's Big Nine, as well as some of George's earliest consulting clients. Over the previous six years, they had done well investing with Oil Drilling. By 1953, the Mitchell brothers were taking a bigger role in the company. They bought out Roxie Wright, splitting the remaining ownership with Merlyn Christie. Merlyn now owned half of Oil Drilling, George owned 40 percent, and Johnny had the remaining 10 percent. Together, they had discovered fields near Palacios on the coast south of Houston, in the Rio Grande Valley, and on the famous King Ranch in South Texas. But they weren't finding much oil. Instead, they hit on a growth formula that was unusual at the time. Larger companies wanted to drill for oil, so, without regard for its name, Oil Drilling began focusing on buying farm-outs for natural gas. Natural gas wasn't worth nearly as much as oil, but focusing on farm-outs reduced the company's exploration expenses, and by taking a smaller share of the deal, George and Johnny kept their drilling costs low. As a result, they could make good money for themselves and their investors.

George thought the idea of a Chicago bookie peddling an oil and gas deal northwest of Fort Worth, Texas, sounded crazy. He agreed to take a look mostly to satisfy his curiosity. The bookmaker, whose last name was Lieberman, turned out to be a middleman shopping the deal in exchange for a small overriding interest. The project covered about three thousand acres, and it had already been pitched all over the country during the past two years.[3] As with Bob Smith's deal in the Panhandle, no one wanted to touch it.

When people outside the state think of the Texas oil business, they usually imagine Hollywood's portrayal of West Texas, where the oil-rich Permian Basin has provided some of the country's largest oil production for decades. Those more familiar with the industry's history may know about Dad Joiner, Spindletop, and the East Texas Field, which gave birth to the oil industry as we know it. But the industry actually started in North Texas almost a decade earlier.

As in other oil-rich regions such as Saudi Arabia, the first drillers in North Texas weren't looking for petroleum deposits; they were looking for water. Texas encompasses almost 269,000 square miles, but all that space contains only one natural lake—Caddo Lake on the Sabine River, which forms the border between Texas and Louisiana. Settlers pushing west relied on rivers, streams, or natural springs for water. Augers and later rotary drilling rigs brought water to the surface when these natural sources couldn't be found. This early drilling activity expanded the understanding of the region's geology. One of these water drillers was sinking a well near Corsicana, southeast of Dallas, in 1894 and hit oil by mistake. A group of businessmen decided to produce the oil and took over the well. It wasn't a gusher like Spindletop, but it introduced the Texas economy to crude. Drilling moved westward across North Texas, and in 1900, legendary rancher W. T. Waggoner, who hunted wolves with Theodore Roosevelt and did business with the last Comanche chief, Quanah Parker, also struck oil while drilling for water on his ranch near Vernon. Waggoner soon realized he was sitting on top of the region's biggest field, which he named for his daughter, Electra.[4] (Electra got an oil field named after her; her daughter, the socialite Electra Waggoner Biggs, sometimes called "Electra II," had a car—the Buick Electra—and a plane—the Lockheed Electra—named in her honor.)

Waggoner's big strike ignited the imagination of ranchers across the region who dreamed that their land, too, might hold great wealth beneath the rolling plains. One of them was Col. D. J. Hughes, a British businessman who became convinced that a river of oil flowed under his 2,800-acre ranch near the community of Boonsville, about forty miles northeast of Fort Worth. To find it, Hughes hired John Jackson, a native of Lufkin, in East Texas, who had a geology degree from the University of Texas. Jackson worked as a high school science teacher, but he learned the oil business working in the East Texas Field in the 1930s before enlisting in the navy after Pearl Harbor.[5] He later moved to Dallas and set up a consulting business. Hughes, known to everyone as Jimmie, told Jackson all he had to do was survey the land and find the oil. Hughes called Jackson each night to learn what he had found.

"What did you find?" he asked.

"Nothing," Jackson replied.

"Keep trying," Hughes said.

Jackson did. Eventually, he drilled a well to a depth of 5,800 feet and struck natural gas, but Hughes wanted oil.

"Someday," Hughes said, "you will find oil deeper. It's down there."[6]

Hughes may not have been interested in natural gas, but Jackson did the math. At the time, it was selling for six cents per thousand cubic feet. The well Jackson drilled, the D. J. Hughes #1, indicated the area contained so much gas that the field could be worth as much as $1 million. Jackson didn't have the money to complete his other two test wells, and he began shopping his deal around to anyone who would listen.[7] In 1951, he pitched it to a stranger he met in the coffee shop of a Lubbock hotel. "He was intrigued," Jackson said, although he never learned the man's name. The lease buyer made a stop in Denver and explained the offer to a friend, who called an acquaintance in Tulsa, who in turn called Lieberman, the Chicago bookie with a side interest in making oil and gas bets.[8] Three days later, Lieberman got in touch with Jackson and asked for more information. Jackson forwarded the well log for the D. J. Hughes #1. Lieberman knew just enough about the oil and gas business that the logs piqued his interest, and he called Pulaski.[9]

By the time the deal came to George, the dreams of another Electra Field had faded. The area had been punched full of so many dry holes it was known to local oilmen as the Frustration Fields. Once again, George was undeterred by what others thought of the geology. The more he studied the maps and logs, the more convinced he became that the others had missed something obvious.[10] George agreed with Jackson's theory that the area contained a major natural gas deposit. Looking at the logs, he saw what he believed was a stratigraphic trap—changes in the rock formations that can retain oil and gas. It also had a large "pay zone," a sand layer containing natural gas that was one hundred feet thick. By the time George saw the well logs, eleven wells had been drilled on the Hughes property—all were dry holes. The Frustration Fields lived up to its name. That alone would have scared off even the most reckless wildcatter, but George noticed something in the well data: all the wells were drilled *through* the gas zone. They basically ran right past what George thought looked like the layer containing the gas.

He and Johnny gathered the Big Nine, Bob Smith, and a few other investors—Sam Maceo had died two years earlier, in 1951—and met with Jackson and his partner Ellison Miles, whom George knew from his college days at Texas A&M.[11] As they stood around a table in Oil Drilling's offices in Houston, studying geologic maps, somebody asked Jackson how big he thought the field could be. He reached for a nearby cowboy hat and dropped it in the center of the maps. "That's how big it is," he said. Everyone stared at the hat covering most of the map. No one spoke at first. Then Bob Smith turned and grabbed Jackson by the arm. "Son, I like your idea. I am going to back you completely—but remember, you had better be right."[12]

With the backing of Smith and the Big Nine, Oil Drilling agreed to invest $10 million to develop the Frustration Fields, which officially became known as the Boonsville Bend.[13] Drilling began later that year, and the first well confirmed George's suspicions about the pay zone, producing commercial quantities of gas. George and Johnny knew they had a big find on their hands, but they didn't celebrate their discovery. Anytime a big well comes in, producers want to keep it quiet. If the word gets out, landowners in the surrounding area will raise the price for their mineral leases. George and Johnny hired dozens of brokers to begin leasing the areas around their initial prospect, known as Hughes Ranch. In three months, they had leased three hundred thousand acres—ten times as much as the original deal—for just three dollars an acre.[14] Most of the landowners were farmers, and in 1953 Texas was baking under the worst drought in its history. Reeling from crop losses and desperate for money, they eagerly took any price they could get for mineral leases on their parched land.

After the first successful well, George and Johnny retraced the steps of the previous drillers, reopening the ten remaining dry holes, which was cheaper than drilling new wells. "The wells were drilled, but they had misunderstood what the formation had in it," George recalled. "They didn't interpret it properly. We moved in [and] we made this big play." By hitting the pay zones that the others had missed, Oil Drilling struck gas in every one of the old wells.[15] George didn't know it yet, but he had made his biggest discovery.

George proved the gas was there, and he, Johnny, and Merlyn Christie had managed to get the rights to drill in a large surrounding area. Despite the complicated geology, George was convinced they had tapped a large gas reservoir, and the more they drilled, the more he seemed to be right. Of the first eighty-seven wells they drilled, eighty-five produced significant quantities of gas.[16]

The price of natural gas was still much lower than that of oil, but George believed inflation would continue and demand for natural gas would increase as oil became more expensive. In the postwar years, natural gas was becoming a fuel for heating, cutting into the market for heating oil, and a preferred fuel for manufacturers such as steel mills. It also had growing uses in making fertilizers and plastics. His views about the overall demand picture, inflation, and his recognition of the emerging market for gas combined into a business strategy. There was only one problem: George didn't have money to fund it. His strategy would rely on large amounts of debt. Debt didn't bother him. He had grown up never having money and scrounging a few bucks where he could. In some ways, he was still the immigrant kid from Galveston hawking fish for pocket change. As he saw it, inflation meant that the money today would be paid back with cheaper money tomorrow.[17]

"He built his philosophy on two facts, or to him they were facts: First there would be inflation as far as the eye could see, and second, the price of gas had to approach equivalency with oil," said Budd Clark, who would join the company a few years later. "With those two ideas, his main thrust was to accumulate as much reserves and land holdings as possible."[18]

The discovery in North Texas would grow slowly, as the company steadily expanded its drilling program, adding more wells in the surrounding counties. But within a few years, Oil Drilling became a much bigger company, and North Texas was on the verge of an economic boom. One visitor, driving up a hill near Decatur and looking across the geologic expanse that's generally referred to as the Fort Worth Basin, thought he was looking at a swarm of fireflies.[19] What he was seeing was lights from the drilling rigs tapping into the natural gas field that few others besides George had believed was there.

◈

The success of the Boonsville Bend development presented a new problem. The Dallas utility Lone Star Gas dominated the local market. Because most producers wanted oil, gas was a consolation prize. If that was all they encountered, the producers would sell it for whatever they could get in hopes of covering their drilling costs. If they couldn't, they flared, or burned it off, and plugged the well. Lone Star used this lack of interest to its advantage and basically dictated purchase terms. It paid a low price—about ten cents per thousand cubic feet—and bought gas only at certain times of year, primarily in the winter when its customers needed it for heating.

George knew the price was too low. He believed that natural gas had more value than the market was giving it at the time. Gradually, others in the industry were coming around to his way of thinking. Still, if Oil Drilling was going to make money consistently from the Boonsville Bend, he needed a guaranteed price that was higher than what Lone Star was paying and, more importantly, someone who would buy the gas year-round. He began looking for a deal that would allow him to circumvent Lone Star's market dominance.[20]

In 1955, two years after Oil Drilling bought into the project, George signed a twenty-year contract to supply gas at fourteen cents per thousand cubic feet to Natural Gas Pipeline Co., which had a pipeline that ran from a gas field in the Texas Panhandle to Chicago.[21] Like George, NGPL needed to be certain of its gas supply. Its reserves in the Panhandle were dwindling, and its market in Chicago was expanding. At the same time, everyone expected both oil and natural gas prices to soar in the coming years. While it was paying above-market prices, NGPL believed it was getting a bargain, and it wanted to secure a steady supply of gas for as far into the future as possible.[22] To do this, it was looking for a big gas field, preferably one close to the reserves it had already tapped in the Panhandle.

Lone Star didn't take the deal lying down. The company tried to convince the locals that Oil Drilling was a bunch of interlopers from Houston who couldn't be trusted. But George and Johnny had hired local land men to acquire mineral rights for the Boonsville Bend in Wise County, near Decatur. The land men knew the local landowners. They went to the same churches and socialized, and because of those connections, Oil Drilling became a major gas producer in the region. To help win over landowners, Oil Drilling threw a big barbecue picnic and unleashed its secret weapon: Johnny Mitchell. The farmers and ranchers couldn't resist Johnny's irrepressible charms. They liked the way he talked and his down-to-earth style, and they listened when he warned them that Lone Star was trying to take advantage of them. "I would rather be my own person and see if we can't get a market for the gas and get a decent price," he told them.[23]

The deal seemed to be done. George had found a long-term buyer who was willing to pay a premium for the gas, and he had leased enough mineral rights to meet the contract demands for decades. The federal government, however, had other ideas. In 1953, the US Supreme Court ruled that the Natural Gas Act, passed twenty years earlier, gave the federal government the right to regulate all aspects of natural gas production. Regulators, reviewing George's deal in North Texas, decided that fourteen cents per thousand cubic feet was too high. They ordered him to cut the price by a penny, which was still three cents more than he would have made by selling to Lone Star. In addition, the feds stipulated that NGPL had to pay for the pipelines and gathering systems to connect the Boonsville Bend wells to its pipeline in the Panhandle. That was critical, because building that system would cost $32 million, and even with Johnny's fund-raising prowess, Oil Drilling would have been hard-pressed to find that kind of capital.[24] Despite the federal interference, George and NGPL agreed the deal was still mutually beneficial. NGPL built a 350-mile pipeline extension from its operations in Fritch, near Amarillo, to the Boonsville Bend. In December 1957, gas began flowing from the field and along the pipeline on a three-and-a-half-day journey to Chicago, where NGPL's quarter million customers used it to heat homes, broil steaks, and make steel.[25] "Keep the gas coming, boys, for we can use all you can give us," NGPL executive Mark Burlingame said during a speech in Wise County in 1966.[26]

There was one other condition. The natural gas coming out of the Boonsville Bend was known as "wet gas" because it contained liquids and condensates that would freeze on the way to Chicago in the winter. NGPL wanted the liquids removed before the gas was put into its main pipeline heading north, and that required a special processing plant.

Neither George nor Johnny—nor Merlyn Christie—knew anything about running a processing plant, so they brought in other partners to take on the project. Bob Smith convinced his bank to lend the company $4.3 million to finance the plant's construction near the town of Bridgeport. It was the beginning of George's lifelong love affair with debt. Although in his later years he would frequently make the *Forbes* magazine list of the richest Americans, he and his company would remain highly leveraged until the early 2000s.

The plant, which is still in operation, proved to be far more than a necessary investment to appease NGPL. It was a crucial piece of infrastructure for the technological revolution that would later unfold in the North Texas gas fields. The gas produced from those early wells was "rich," meaning that it could be processed to strip out ethane, butane, and propane, which Oil Drilling would sell to chemical companies on the Gulf Coast. Larger companies at the time generally ignored these niche markets for such by-products.

The plant began processing gas from the Boonsville Bend wells in 1957.[27] In time, George would expand the technology and operate dozens of similar plants. He realized that the seemingly marginal products being stripped out of the gas were increasingly important to chemical companies that were producing growing quantities of plastics.[28]

The liquids business, though, was a mere shadow of the technological breakthrough to come. George didn't know it yet, but in landing the NGPL agreement, he had signed one of the most important contracts in the history of the energy business. It was the kind he could build a company around, and he did. NGPL would continue to buy gas at above-market prices until 1995. But more importantly, the need to meet the production requirements of the contract in its final years would create the incentive for the development of fracking.

CHAPTER 8

Beyond the Big Nine

As the 1950s progressed and the Boonsville Bend came closer to production, George settled into his success. Far from ending up a bum, as Cynthia had worried, George found that going into business with Johnny laid the foundation for his prosperity. He now made enough money that his growing family didn't have to struggle. In 1950, when they moved into the four-bedroom house south of downtown Houston, George was just thirty-one years old and Cynthia was twenty-eight, yet they had already achieved a living standard that neither had known growing up. Eight years later, when they left the house, they were comfortably raising eight children—three girls and five boys. In addition, Cynthia's sister would drop her three children off during the day.

The Mitchells weren't done adding to their family, and the size of their brood often surprised George's associates. He hung their pictures in a line on the wall of his office. When Shaker Khayatt, who would become a director of the company, first saw the string of photos, he said, "George, that's very interesting: I never thought of taking pictures of my children at different ages."

George shot back: "What are you talking about?"

"Well," Khayatt said, "you have them young here, and then about five years later."

"Wake up," George replied. "They're all different."[1]

Fatherhood didn't come naturally to George, and he wasn't present for his children's births. But he also seemed determined to provide a sense of stability for his children that his own childhood had lacked. He loved his family unconditionally, even if he could be quiet and distant at times.

Most afternoons, George would leave work for his regular tennis match, have a beer with the other players, and get home at about 7:00 p.m. The children had already been fed by the housekeeper, and she left dinner in the warmer for George and Cynthia, who didn't enjoy cooking. Some nights, George would grill dinner for Cynthia, but the ritual of tennis and the late arrival annoyed her.

Cynthia wasn't a typical 1950s housewife. She was college educated, and while she certainly enjoyed George's business success, she viewed their marriage as a mutual partnership. Like George, she had to work at being a parent, and she often complained that she felt misunderstood or unappreciated for her intellect and creativity. At the same time, she served as the backbone of the family, stressing art, language, compassion, and education for the children.

She loved the arts and sciences, particularly cosmology—in later years, she faithfully watched astronomer Carl Sagan's show *Cosmos*, and she could perfectly imitate his catchphrase, "billions and billions." She loved words and literature, too, writing scores of poems that were never published. She would correct her children when they mispronounced foreign words, and if they asked her what a word meant, she would refer them to the unabridged dictionary she kept in the study with the command "Look it up!" As her children grew older, she would often stay up all night with them proofreading and typing their term papers.

She refused to be defined solely as the wife of an oilman. "She didn't buy into the successful tycoon role," said her daughter Sheridan Lorenz.[2] Although George was gaining prominence as a businessman in Houston, he and Cynthia had little interest in joining the city's social elite—most of whom had also made their money from oil. Cynthia, in particular, was uncomfortable with being in the society pages, and she once wrote to Maxine Mesinger, the *Houston Chronicle*'s society columnist, and asked her not to mention the Mitchell family in her column. Cynthia worried that the publicity would affect her children, and that being featured in the paper would draw the attention of burglars or kidnappers. Later, of course—after her name was on an outdoor theater, she and George had led the restoration of Galveston, and The Woodlands had opened—publicity was unavoidable. But even then, the articles focused on achievements that benefited others. She had little time for the gossip columns detailing who ate where for lunch or what designer gown someone wore at a gala.

While Cynthia tended to push the children to succeed, George was more hands off, as if he just expected them to do well in school or other endeavors. At the same time, he was patient and even indulgent. If one of kids' lines got tangled during a fishing trip in Galveston, he would set down his own rod and patiently pull apart the knots, even if it meant he didn't get to fish himself. "He almost got some weird pleasure out of untangling these incredibly tangled up fishing lines," Sheridan said. Fishing, in fact, became a regular family pastime. George was often happiest when he was wetting a line—the only leisure activity he loved more than tennis. He rarely fished with coworkers or other oilmen, although he would sometimes go out with childhood buddies in Galveston. He found a degree of comfort in taking the children, perhaps because it didn't involve any demands for social niceties or small talk. He knew every jetty and fishing spot, including his favorite, Redfish Hole. Cynthia made a chart for the back of the door to track which children would go when—and made sure the girls took as many trips as the boys. Most summer mornings, when the family typically was in Galveston, George would take the kids for a fishing trip before heading to work. In later years, he often took the children for weekend trips to Cook's Branch, a family retreat north of Houston. He loved to build a big fire in the old log cabin on the property in the evenings, and in the morning, he would fix pancakes for the kids. During the week, he regularly dropped his older children off at school before heading to work.

George was not a jet-setting oil executive. He hated traveling, and he didn't understand other people's love of vacations. Once, when his children were grown, he tried calling all ten at different times and couldn't reach any of them. When he finally got hold of one of the boys, his son pointed out that it was the middle of summer, when most people take vacation.[3] George didn't seem to understand.

When some of the children's activities caught his interest, though, George could be as focused on them as he was on work. He organized regular family stargazing nights, taking all the children to the beach on Galveston Island and teaching them about the constellations and planets. He loved contemplating the night sky, as he had done since childhood. "I always thought that was his religion of sorts," his son Mark said.[4] When two of the older boys started building a kayak in the garage, George became fascinated and insisted on finishing it long after they had lost interest. His children were one of the few priorities that superseded work and tennis. Anytime one of them called his office, he always picked up the phone, regardless of what he was doing or whom he was meeting with. And even though he often worked on weekends, he rarely traveled on Saturdays or Sundays, so he would have time for family activities.[5] In later years, when his son Kirk became seriously ill in Brazil, George canceled a company board meeting, flew to the country, and spent a week at his hospital bed.[6] It was as if, after the broken home of his childhood, he was determined to keep his own family together no matter what.

Neither Cynthia nor George spanked the children, even though corporal punishment was still in vogue. George, in particular, avoided conflict. Cynthia was more discerning as the children grew older and put more demands on them. She wanted her children to contribute to society and to follow their passions, and she made her expectations known. "I'm not just trying to raise children who don't go to jail," she would say. George, on the other hand, tried to create an atmosphere of trust. Often, he would seem to believe things his children told him even if he knew they weren't true. If he was forced to confront a lie from one of them, he would grow visibly upset—not angry, but saddened, as if they had betrayed him. Just as in business, where he preferred handshake deals and trusted his employees implicitly, George wanted to trust his children.[7] "He wanted to believe in you so much," his youngest son, Grant, said. "He was loyal to a fault."[8]

Together, George and Cynthia were the epitome of postwar success. Neither had come from a traditional home. George had grown up without a mother amid a raucous gaggle of Greeks, and Cynthia had been raised fatherless by subdued midwestern farmers. Both had had to help support their families and themselves at an early age, and both had learned to overcome those hardships and succeed in college. Neither really knew what a two-parent, middle-class family was like, yet they managed to do all the things families did. The kids were involved in sports and other activities, and the Mitchells went to the Episcopal church on Sundays. They were active in the congregation, although neither George nor Cynthia was particularly religious. George was raised Greek Orthodox, and Cynthia was Roman Catholic, but they became Episcopalians in 1949. Several of the children received perfect Sunday school attendance awards, and the Mitchells hosted pool parties for other church schoolchildren. Cynthia ran the vacation Bible school and was active in the women's group. George was less involved, but he appears in several Easter photos from the 1950s with his children.[9]

The couple adored each other, despite George's occasional aloofness. He respected Cynthia's creativity, intelligence, and generosity—she was known for pulling up with a station wagon full of presents to donate to children's charities.[10]

By 1953, with the company growing and the success of the Boonsville Bend looming, George decided he wanted to share his good fortune, starting with his alma mater, Texas A&M. Decades later, he would become the school's biggest benefactor, but he started small. The Petroleum Engineering Club's tradition of awarding a gold watch to the best student fizzled a few years after George graduated. No one was getting watches like the one he'd flashed to impress Cynthia and her sister on the train back in 1941. He decided to restart the tradition, buying inscribed gold watches for the senior and junior with the best grades—a tradition he would continue for sixty years.[11]

His family, too, felt the benefits of Oil Drilling's rise. He and Cynthia built a weekend home in Galveston. Cynthia, whose love of art and design hadn't waned since her days at the University of Houston, hired renowned Texas architect O'Neil Ford to design it. Ford had a modernistic style that appealed to the young couple, but he had two other qualities that would define the Mitchells' views in years to come. He was a strong preservationist who taught other Texas architects to look to the state's frontier past for inspiration in design and materials, and he designed homes that were intimately tied to their surroundings.[12]

The family made regular weekend trips to Galveston, where Mike

Mitchell continued his carousing and George's brother Christie ran his bar and drove around the island in an old Cadillac convertible, which he called the Blue Goose. With his ever-present Panama hat and cigar, Christie was a larger-than-life figure.[13] He and his wife, Janie, were well-known ballroom dancers, and when they took to the floor, everyone would stand aside to watch them. (Christie once won a dance contest with Sam Maceo's wife, beating out professional dancers from an Arthur Murray studio in Houston. The prize was a stay at the Buccaneer Hotel in Galveston, where, ironically, the Maceos lived.)[14]

Christie wrote his newspaper column first under the pen name "The Night Owl" and later as "The Beachcomber" (which was also the name of his bar). In one column, he claimed that a girl he had dated in college, who was part gypsy, had predicted that Mike's beloved Snug Harbor Cafe would catch fire.[15] It did, in July 1947, under mysterious circumstances that echoed those of the Chiacos & Mitchell fire in Houston in the early 1900s. This time, no lives were lost, but Mike was devastated. He was sixty-seven years old by then and had turned operations of the café and the Neat Dresser's Club over to others, but he, George, Johnny, Christie, and Maria owned the building. According to Mike, the fire started after a delivery of naphtha, a flammable liquid, to the pressing shop. Some spilled on the floor and was ignited by a blower that was knocked off the furnace.[16] "He took it pretty hard at first," Christie wrote. "But being a Greek philosopher, he recovered." Mike moved in with Christie at the Coronado Courts apartments, where he continued to run up big bills buying nice suits, expensive hats, and stockings for various lady friends.[17]

While George disliked straying too far from his beloved Galveston, Cynthia would occasionally convince him to take short trips. She loved snow, and she convinced him to try skiing in Stowe, Vermont. George hated the cold, but he enjoyed skiing. Vermont was too far from Texas for regular trips, so in 1956, he and Cynthia took their older children to Aspen. Beginning in 1958, the family would go every year until the late 1990s, sometimes taking the train. As an island native, George had a steep learning curve when it came to skiing, but he attacked it with his usual competitive zeal.

Riding the lift one day, he sat next to a visitor from overseas, and the two men began talking casually. George asked why the man had come so far just to ski, and the man said that Aspen had some of the world's best slopes. George had been thinking about what Bob Smith had told him about pursuing a second fortune on top of—rather than inside—the land. By the time he reached the end of his run, he decided that if people were coming from around the world to ski in the sleepy old mining town, he had an opportunity to put Smith's advice into action. Within a few years, George had bought all the land at the base of Aspen Mountain, using the site for a family home, where they would spend each Christmas, and later a condo development, the Aspen Alps.[18]

◈

George and Johnny were more than just the perfect pairing for doing oil deals; they also represented the old and new philosophy for the business. Johnny was the gambling money man, the kind of prospector who'd been drawn to the business from its earliest days. "When you gamble with nature, it's like a love affair," Johnny said. But it no longer was. To Johnny, the business had become cold and calculated. George's approach—scientific and analytical—was the future. "He's a new type of breed we didn't have in the old days," Johnny said. "And he doesn't have any fun. I've had more fun in five years than George's had in ten."[19]

As drilling costs increased, wildcatters couldn't afford their fun. No longer could they just punch holes in the ground based on gut instinct. As the industry grew, it became increasingly clear: Dad Joiner had been one of the luckiest men of all time—and luck isn't a business model. As the demand for oil and gas rose, the oil business was growing up. They couldn't afford the freewheeling, roll-the-dice ways of a Glenn McCarthy or a Roxie Wright anymore.

"We're more scientific than the old wildcatters you've heard about," George said. "A real wildcatter thinks he's one foot from a million dollars, when in all reality, he's a thousand feet from one dollar. You can't be a wild optimist, because a wild wildcatter—you'll find that most of them never make it."[20]

Oil Drilling was no longer a ragtag little company. It owned producing properties in South Texas, it had opened a field office in Canada, and, of course, it controlled one of the largest gas discoveries in the country, the Boonsville Bend.[21] The growth from that project alone required greater corporate maturity. Just before the NGPL pipeline began transporting gas from the Frustration Fields to Chicago in 1957, George, Johnny, and Merlyn Christie created a holding company for Oil Drilling: Christie, Mitchell & Mitchell, or CM&M. It added a layer of complexity to what had been a straightforward corporate structure, but it served an important legal purpose. CM&M was the operator of all of Oil Drilling's wells, which meant it handled the drilling on behalf of Oil Drilling's investors. The new structure shielded those investors and the partners from any liability from the operations. For the same reason, CM&M itself had few assets.[22]

The next stage of the overhaul came a year later, when Merlyn Christie hired Bernard F. "Budd" Clark, a chemist and native New Yorker who got his Harvard MBA after serving in the Army Air Corps. He worked for the New Orleans oil company Pan Am Southern until it went out of business, and then, as part of his job search, he sent his résumé to the Harvard Business School Alumni Association in Boston as well as its chapter in Houston.

About the same time, Merlyn Christie decided the company needed some professionalism in the management structure, and as he saw it, the best way to do that was to hire a Harvard Business School graduate. How he found Clark remains unclear. Merlyn himself claimed that he was at a bar in the Ambassador Hotel in New York and simply asked a man drinking next to him how to find one. The man gave him sage advice: call the Harvard Business School. But as Clark noted, Merlyn "was never constrained by details."[23] It's more likely that Clark's résumé made its way from the Houston chapter of the alumni group to a vice president at First City Bank in Houston, who in turn gave it to Merlyn. Clark, who died in 2005 at age eighty-four, never knew which version was true.[24] All he knew was that Merlyn Christie called and asked him to come to Houston.

Clark told Merlyn that Conoco had also invited him to town for an interview a few weeks later, and he intended to combine the trips. Merlyn grew impatient. "Do you want the job or don't you?" he demanded. Clark agreed to come sooner, and Merlyn hired him before Clark ever met with Conoco.[25]

Clark didn't handle operations. Merlyn hired him to clean up the accounting. George, Johnny, and Merlyn hadn't kept up with the finances as the business had grown. Audit reports and payments to investors were often late by three months or more. Clark implemented an estimated billing system to get the money coming in sooner, a critical improvement because banks wouldn't lend against an unsuccessful well. They wanted to see income or other proof the company could pay its bills.[26]

By then, the development of the Boonsville Bend was in full swing. The company had drilled hundreds of wells, but the drilling program required a constant flow of capital to keep the project going. CM&M was becoming a victim of its own success. "Everything was going great and it was profitable, but the need for capital was just limitless," Clark recalled. For the Big Nine, investing in the North Texas field was little more than a hobby. They had no interest in funding a major drilling program. One by one, they each told George that they wanted to maintain their interest, but they wouldn't make any additional investments.[27]

Johnny may have felt the business moving away from him, but he hadn't lost any of his money-raising skills. He drummed up a steady stream of new investors. George's reputation for finding oil opened new doors to investment circles in the Southwest, and Johnny brought in well-heeled backers like oilman John Riddell, Woolworth heiress Barbara Hutton, and Stephen Clark, heir to the Singer Sewing Machine Co.[28]

The deals became increasingly intricate, because none of the investors could afford to keep pumping money into the project at the rate CM&M needed. For years, the company had operated the same way it did when Oil Drilling first started—it raised money from investors, drilled a well, and then raced to the bank when the well came in to borrow enough to complete it. It was a risky, shoestring approach to financing, and CM&M now needed more reliable funding. George's petroleum engineering professor Harold Vance had left academia to run the energy department at Bank of the

Southwest in Houston. Vance studied the logs from the Boonsville Bend and saw the field's potential. "He did something that was done rarely, if at all, in that he lent George money on the basis of the logs showing what reserves were behind each well, even though the well wasn't producing," Clark said. George also convinced Natural Gas Pipeline to advance money for production payments. Altogether, CM&M raised about $7.5 million.[29]

The entire North Texas operation—the drilling, the leasing, the processing plant—had an insatiable appetite for capital. Even some of the newer investors Johnny lined up began balking at demands for additional funding. In late 1958, Oil Drilling got a bit of a reprieve. Institutions as well as individuals began taking notice. Northern Illinois Gas Company and an investment group put together by Empire State Bank in New York pumped more than $2 million into the operation, and CM&M used the money to drill fifty new wells in a month's time.[30]

But as the expansion went on, Riddell, Hutton, and Clark began questioning the pace of the drilling and the rate at which CM&M burned through cash. They also wondered whether the project was managed properly, and they hired an auditor who spent a year scouring CM&M's books. Thanks to Budd Clark, the auditor found everything in order, but the new investors weren't satisfied. They wanted to buy out George, Johnny, and Merlyn Christie and bring in their own management team.

The constant need for capital and the relentless demands from investors began to wear on George and Johnny. "Hell, even when we had 200 wells a year and we were up there with the top 10 independent [oil producers] the investors were calling every Sunday giving us hell wanting to know what we were doing with *their* money," Johnny complained.[31]

George wasn't about to give up on the Boonsville Bend just as it was nearing fruition. He knew his North Texas property was worth far more than any of his investors—except Bob Smith—realized at the time. They would be walking away from a potential fortune. As he would do so many times in the future, George valued the gas field in North Texas not for its current worth, but for its long-term potential.

Johnny, ever the deal maker, devised a plan to keep the company.

He had moved beyond his days of backslapping at the Esperson Building coffee shop, and he had built an extensive network of wealthy individuals, bankers, and other money men. He convinced a contact at Chase Manhattan Bank to provide enough financing so CM&M could turn the tables on the investors. Instead of losing the company, George, Johnny, and Merlyn Christie bought out Riddell, Hutton, and Clark.

They had managed to hold on, but George had had enough. He was done sharing ownership in deals with uncooperative partners. He bristled at their shortsightedness. He still needed to raise capital, but from then on, he would keep control of the projects he and Johnny put together. And slowly, he would begin to consolidate control of the company.

CHAPTER 9

Growing Concerns

George had turned the Frustration Fields into a financial engine that would continue to drive the company for decades. He had also achieved the measure of personal success that Harold Vance had told him would come with striking out on his own—a Cadillac. He drove the same model sedan for the rest of his life, but rarely were they shiny and new. His first one was old and pink, which was a popular color for the car in the 1950s.[1]

In 1958, George moved his family to a larger house in west Houston. The neighborhood was among the newest and most desirable in the city, but George soon noticed that modern developments lacked the lush greenery that he had come to appreciate around the old house on North MacGregor. "There's no trees," he commented to one of his employees. "People like to live in the trees."[2]

The company was on the move, too. George became president of Christie, Mitchell & Mitchell in 1959, and Merlyn Christie became chair.[3] A few years earlier, in the spring of 1955, the firm had moved into offices on the twelfth floor of the eighteen-story Houston Club Building downtown. Established in 1894, the club had become *the* social gathering spot for the powerful elite in Houston's business and civic communities by the time CM&M moved into the building.[4] George would maintain his membership there—and frequently have business lunches in its dining room—well into the next century, long after the club's glory had faded. The offices themselves were functional but not fancy. Visitors entered a small reception area with two switchboard operators. The receptionist's desk, which was also squeezed into the little room, remained empty. George had the phone operators greet visitors and handle deliveries. His office was more spacious, but the decor was already dated when he arrived, and he never refurbished it during the decade or so that the company remained in the building. Joyce Gay, George's assistant, had a tiny reception area, and Budd Clark's office down the hall wasn't much bigger. All the executives were stationed along one wall, with George at the end. Employees referred to it as "King's Row."[5] "There wasn't anything flashy about those offices," Gay said. "They just didn't think that way in the beginning."[6]

Although he was the liaison to the banks and held the title of chair, Merlyn Christie was largely uninvolved in daily operations by the early 1960s. He was a trader by training, and his personality could be abrasive, often rubbing employees the wrong way. He and George remained friendly, but Merlyn believed the company was growing too quickly, and he decided to retire. In late 1962, George and Johnny bought out Merlyn's interest for $16 million, changing the name of the company to simply Mitchell & Mitchell Oil and Gas Corp.[7] Soon after, they also bought out Bob Smith. Two years later, the company was worth about $70 million, employed about 350 people, and had interests in a thousand oil and gas wells.[8] George and Johnny owned the entire company—80 percent for George and 20 percent for Johnny. The reason for the disparity was simple: "George likes to make money, and I'm glad he does," Johnny said. "We agreed that Georgie would get the [80] percent because he wanted to keep working."[9]

Indeed, one year later, Johnny, too, decided to leave the company. He liked the idea of finally having some money, but he lacked his younger brother's interest in building an ever-larger enterprise. The move put George exclusively in charge, and he soon changed the name of the company to George Mitchell & Associates. The early investors in Oil Drilling, CM&M, and Mitchell & Mitchell had done very well. "Any investor who stayed with us for three years made money," George said in the early 1980s. "We had our bad years, but over a three-year span, we always made money."[10]

Johnny, of course, didn't really leave. He started doing oil deals on his own again, and he set up in a small office elsewhere in the building. He frequently wandered into the GM&A offices to borrow the expertise of George's geologists and engineers. "Johnny was never very far away," said Howard Kiatta, who joined the company in 1967. He recalled a meeting in which Johnny came and asked him to review some geology for a deal he was working on. George barged into Kiatta's office and impatiently began asking questions about a different deal that Kiatta was involved with. George wasn't wearing a coat, but he had on a tie, dress shirt, and suit pants—his typical work attire. One of the pockets of his pants was torn. Johnny said, "George, you look like crap, why don't you buy a new suit and some new clothes?" Without looking up, George shot back: "Johnny, if you'd pay me some of the money you owe me, I might be able to afford some new clothes." Then, he immediately returned to peppering Kiatta with questions. "Johnny was always pretty dapper and George, although he always looked good, he never was concerned about being as dapper as Johnny was," Kiatta said.[11]

Johnny may have been a snappier dresser, but he once again struggled in the oil business without George's help. Even before George bought his brother's interest in CM&M, Johnny became president of a small, publicly traded drilling company in Los Angeles, Jade Oil. The outfit drilled in places like Long Beach and Santa Monica and ran newspaper ads with glaring headlines promising good leasing terms for mineral rights. But Johnny's luck hadn't changed since his days in Louisiana. Jade lost money, racked up big debts, and ran afoul of the Securities and Exchange Commission for selling unregistered stock. Angry investors called so often that the secretary learned not to say the name of the company when she answered the phone.[12] Eventually Jade filed for bankruptcy.

Publicly, George distanced himself from Johnny's failed venture. When Jade went bankrupt, he sent a memo to his employees noting that Jade belonged to Johnny and had no connection to George Mitchell & Associates. None of his employees needed to worry about the filing. While it was technically true that Jade had no ties to GM&A, they did business together.[13] Jade acquired interests in some of GM&A's Boonsville Bend wells.[14] What's more, George personally owned 8 percent of Jade's stock, and Johnny owned 15 percent.[15] Some GM&A employees received Jade stock options as part of their annual incentive packages. In one case, Budd Clark called a young geologist who'd only recently joined the company into his office. According to the geologist, who asked that I not use his name, Clark asked him: "When was the last time you took a vacation?" The employee said he planned to take one with his wife later in the year. Clark told him he should take one sooner. The geologist said he really wasn't interested in doing that. Clark pushed an envelope with the employee's Jade stock options across the desk and again strongly urged the geologist to take a vacation soon, but the geologist again brushed off the suggestion. Clark sighed and finally said, "Jade is filing for bankruptcy tomorrow." The point was clear: exercise the options now and take the vacation so it doesn't look like you're trading on inside information.

◈

As Johnny was winding down his involvement in GM&A, he had become an increasingly vocal spokesperson for the struggles of independent oilmen. In 1964, *Businessweek* said that the former artillery officer's "fusillades now boom with peppery regularity through the dwindling ranks of the independent oil industry."[16]

By the early 1960s, independent producers like GM&A were struggling. They had largely created the modern oil industry, overcoming attempts by dominant leaders like John Rockefeller to consolidate control of the domestic market, and discoveries like George's in North Texas showed how independents were willing to take risks that bigger oil companies wouldn't. Because they were smaller and had less overhead, independents could make money on lower-production wells. They didn't need every discovery to be a major gusher. During the industry's early years, the bigger companies—mostly the remnants of Rockefellers' Standard Oil Trust—tolerated their smaller competitors. But over time, they used their lobbying power to enact production controls that worked against their smaller rivals.

After World War II, the major oil companies pursued oil reserves on a global scale. In February 1945, US president Franklin Roosevelt left the gathering of Allied leaders at the Yalta Conference and secretly flew to Egypt, boarded the *USS Quincy*, and steamed up the Nile to the Great Bitter Lake. On Valentine's Day, King Abdul Aziz, who had united modern-day Saudi Arabia, was airlifted onto the deck of the *Quincy*. The king, who was known by his royal title Ibn Saud, rarely traveled outside his kingdom, but he had a pressing issue on his mind: persuading the American president against creating a Jewish homeland in Palestine after the war. He and FDR, both wheelchair bound, talked for the next five hours. While no transcript of the discussion exists, it's likely the two leaders, who developed a fast friendship during their meeting, discussed oil in addition to broader issues of postwar politics. Only a few years earlier, American oil companies had made significant discoveries in Saudi Arabia to help fuel the war effort.

FDR died two months later, in April 1945. King Abdul died in 1953, inciting a power struggle between his two sons, Saud and Faisal. Faisal eventually prevailed, and he took an active role in developing the Saudi oil industry as a means of building the nation. The goodwill that had blossomed between FDR and King Abdul aboard the *Quincy* gave rise to a long-standing friendship between the United States and the Saudi kingdom. American oil companies went on to play a major role in Middle East oil production.[17]

In the United States, the domestic energy market was changing dramatically. By 1948, just three years after the meeting on the *Quincy*, the nation no longer exported more oil than it consumed. In the next few years, it became increasingly clear to independents like GM&A that they would be marginalized as cheap imports from Saudi Arabia supplanted their domestic production. At first, the effects were minimal. The Marshall Plan governing the rebuilding of Europe after the war sent most Arabian crude exports to the Continent, allowing US producers to continue feeding their home market. But one thing was clear: America was no longer the world's biggest supplier of oil.

Meanwhile, the Texas Railroad Commission—which, as the regulator in the country's biggest oil-producing state, essentially controlled US oil production—cited declining demand and lowered the allowable levels of production. It had adopted the allowables system decades earlier to maintain a higher, and thus profitable, price of oil. By 1957, the number of drilling rigs working in the state had fallen by more than five hundred from just a decade earlier. Fewer wildcatters like GM&A searched for oil. What was the point? Even if they found it, they probably couldn't produce it because of the commission's production caps. No one wanted to buy domestic oil when imports were cheaper. The industry called for Congress to enact import quotas on Saudi oil, but the lobbying and public relations efforts of the major oil companies intervened. Imports, they argued, were good for the American public because they kept gasoline prices low, and any damage to the domestic oil industry was temporary. Besides, the independents were a bunch of swashbucklers and loudmouths who didn't like the idea of competition. The majors, though, had a huge financial interest in imports, and they convinced Congress to adopt a tax scheme that essentially meant that as their profits from imports rose, their taxes went down. Domestic producers, still paying taxes that totaled about seventy-five cents a barrel, simply couldn't compete.

Policies favoring the majors didn't stop with oil. In 1953, the US Supreme Court decision that had forced George to cut the price he negotiated with NGPL in North Texas also gave the federal government the right to regulate all aspects of natural gas production and treat producers who sold gas across state lines as utilities. By 1962, these policies had established a clear trend: rising oil imports and declining production in states like Texas.[18] The number of independent oilmen like the Mitchells declined, too. One of their contemporaries, Michel Halbouty, estimated that the number of independents nationwide fell to 12,000 in 1964 from 30,000 in 1957—a 60 percent decline in seven years.[19]

Independents struggled to keep pace with these industry changes. As imports grew cheaper, domestic drilling became more expensive. To keep finding new oil, producers had to go deeper, which meant paying for more drill pipe.

In 1963, a year before he left CM&M, Johnny found himself in the role of industry lobbyist. He became an outspoken voice firing off letters to lawmakers warning that the ranks of independent oil companies were dwindling, and with it, our nation's energy security. In December 1963, almost a decade before the first Arab oil embargo, Johnny wrote to the chair of the multinational Standard Oil Co. of New Jersey (which later became Exxon): "The day will come when all the [oil-producing nations of the Middle East] band together . . . turn the valves of every oil well to the right and watch the oil-hungry free world, including our invincible, resourceful United States, fall to its knees, subdued and begging."[20]

Congress wanted to tax oil and gas property sales as ordinary income rather than capital gains, which, many independents feared, would reduce the amount of acreage for sale and drive up land prices. Johnny flew to Washington to meet with Wilbur Mills, chair of the House Ways and Means Committee. Armed with a scratch pad and pen, he scribbled out his argument that the move would put independents out of business. Mills later used Johnny's notes to convince the committee to drop the measure.[21] Although Johnny succeeded, the situation underscored a larger problem for independents: public perception. Wildcatters' independent nature meant they didn't like asking for help, and because they didn't have direct contact with the public, politicians showed little concern for their plight outside oil-producing states like Texas and Louisiana. "We've never bothered to talk to consumers," Johnny groused. "We've always gone to Washington when the house was burning down looking for a fire extinguisher, too goddamned late. We are to blame for most of our problems."[22] Those comments also presaged the public outcry over fracking more than three decades later, when once again, the industry ignored public concerns until it was too late.

If he wasn't twisting arms in person, Johnny was writing letters. He fired off dozens of missives to politicians, federal and state regulators, other independents, and executives at major oil companies. He sent copies of every letter he wrote to newspapers as well. Almost all of them had the same theme: the majors wanted to put the independents out of business. He mixed in a bit of flag waving, too. The majors, after all, were abandoning America to seek oil overseas. The independents, on the other hand, were still finding oil at home.[23] In 1979, Johnny would compile the letters—almost one hundred of them—into a book titled *Energy—And How We Lost It*. It chronicled how the majors "decided to flood the American crude market . . . with oil which supplanted American production," according to an introduction by fellow oilman James Clark. The influx of imports sent prices down, crushing independent producers. "Gleeful, but short-sighted consumers cheered and consumed their way directly toward a disaster"—the energy crises of the 1970s. "It was all too late. The domestic oil industry was on its downhill slide."[24] Not everyone agreed with Johnny's bleak message, of course. "We don't accept him as a spokesman," one unnamed oilman told *Businessweek*.[25]

George shared his brother's concerns about the changes he saw in the business, but he didn't have Johnny's bombast. Johnny penned several screeds blasting Texas senator Lyndon Johnson, then the state's most powerful politician, for his energy policies. Johnson's aides called George to complain. "If you want my brother," George replied, "I'll sell him to you."[26] George favored an import tax on oil, but he had little patience for most politicians. They paid too little attention to energy policy and the role it played in national security. "I believe that the oil and gas industry deserves more attention from the government than other industries because it is much more important than steel and autos and others," he said. He also bristled at how politicians, when they did pay attention to energy, focused on oil and overlooked the benefits of natural gas. In prescient comments made in 2000, he declared that America's dependence on foreign oil could be reduced in a decade by developing more natural gas, which, he said, "is a better fuel than anything else. Natural gas has fewer [carbon dioxide] problems than other fuels and there's no acid rain to deal with."[27]

But, as George saw it, the country was squandering the opportunity that natural gas presented. He blamed poorly conceived federal regulations for limiting the price of gas and creating market uncertainties that kept companies from drilling more wells. "Under the present price structure, there is not enough risk capital going into gas exploration," George told an industry conference in Minneapolis in 1965. "Much of the capital of major companies has fled abroad in tremendous sums to finance the worldwide exploration for oil the Mideast, Libya, Algeria, [the] North Sea, Nigeria and other places. The capital is not available to explore in this country because the market price and reserves of natural gas and oil in the US are not as attractive as abroad."[28]

Rather than badgering congressmen, George began looking inside the company for ways it could protect itself from the changes that were coming. GM&A needed tighter internal controls. If independents were to survive, they had to act more like real businesses and less like gamblers. He also noticed something else as he studied the shift in the industry: the majors didn't plow all their profits back into exploration. They diversified into other interests. In the early 1960s, the trend in business was to form conglomerates. Some of the biggest US companies acquired disparate holdings. Dallas-based Ling-Temco-Vought sold stereos and jet fighters; AMF Corp. started out making cigarette machines and after World War II produced bowling equipment and nuclear reactors. Tandy Corp., a maker of leather goods, bought the RadioShack electronics chain and the flooring company Color Tile.

George wasn't considering anything so drastic, but as the threat from imports drove down domestic energy prices, he felt increasing pressure to diversify the company. In a move that foreshadowed his later interest in real estate, he acquired land on Galveston's west end and built two developments, Pirates Beach and Pirates Cove, both named after the notorious pirate Jean Lafitte, who used the island as a base in the 1800s. George and Johnny also each bought 30 percent interests in Pacesetter Homes, giving them control of the company. By 1964, Pacesetter owned seven subdivisions, three parks, and an office building in Houston.[29] Pacesetter, though, never made any money, and Pirates Beach and Cove were little more than individual housing developments clustered around a golf course.[30] They bore little resemblance to the master-planned masterpiece George would build later.

Perhaps because of Bob Smith's advice about building a second fortune *on* the land, George liked using real estate as a hedge against downturns in the oil business. He told the freelance brokers he hired to find oil and gas leases that they should keep an eye out for good real estate properties. In 1962, GM&A made two major purchases, both fifty thousand acres—one on Pelican Island, near Galveston, and the other in southern Montgomery County, north of

Houston. Pelican Island included three thousand acres of mostly industrialized land in the middle of the bay between Galveston and the mainland. George donated one hundred acres to his alma mater, Texas A&M. The university used it to set up a Galveston campus focused on marine sciences, which George named for his father and mother, who had never gone to college. For the forestland north of Houston, though, he had few immediate plans. He'd read enough studies forecasting the Houston area's growth to know the city's population was moving northward, but he wasn't sure what, if anything, he would do to capitalize on it.[31]

The Grogan Cochran Lumber Co., a struggling timber company controlled by three local families, had owned the land since 1903.[32] By the early 1960s, disagreements among the ninety heirs had stymied the firm and drained its finances. "They couldn't even buy a pack of cigarettes they were so broke," George recalled in a later interview.[33] He liked the property, with its beautiful swaths of dense pine forests between the San Jacinto River and Spring Creek.

George met with the various factions for a year and a half, finally hammering out a deal to buy the property for $125 an acre, or about $6 million. He didn't have the cash, but just as he had in the oil business, George learned to get what he wanted with other people's money. He convinced his old professor Harold Vance at Bank of the Southwest to advance him the down payment. He used the funds to get a release on part of the land and then went back to the bank and used that land as collateral for the loan. He struck a deal with the lumber company Georgia Pacific to harvest timber from the property, which generated about $5 million.[34] Then, he sold the timber contract to an insurance company and used the proceeds to repay the bank loan. He had managed to finance the purchase of fifty thousand acres with almost no money of his own.

George had one stipulation for the deal, and it involved the timber contract: no clear-cutting. The contract called for harvesting using what's known as a "real estate" or 12–12 cut—trees were cut 12 inches above the ground, and they had to be at least 12 inches in diameter. That allows the forest to regrow more quickly, because it spares most of the new timber growth.[35]

The timberland wasn't contiguous but scattered throughout the western part of the county. One tract, though, caught his attention. It was about 2,800 acres close to Interstate 45, the main north-south conduit through the Houston area. If you took the freeway south, it ran directly through the center of Houston and then kept going south until it ended across the causeway in George's native Galveston. The land had lots of potential, thick with towering loblolly pines and choked with yaupon and dense underbrush. George wanted to do something more than just harvest timber. But what? He pondered the question: "What could we do?"[36]

CHAPTER 10

Social Awakening

As George diversified his business interests, he also expanded his awareness of social problems. A voracious reader since working his way through the Galveston library as a child, he had an intellectual curiosity that generated a lengthy and wide-ranging reading list. As an adult, he rarely read fiction. Rather, he devoured books and essays on science, nature, and deeper philosophical issues facing society, particularly population growth. By 1963, he had fathered ten children, and he became increasingly concerned about the sort of world his heirs would inhabit after he and Cynthia attended a lecture by inventor and futurist Buckminster Fuller at the Aspen Institute in 1959.[1]

Fuller envisioned new ways of building homes and cars. He's best known for inventing the geodesic dome, designed to enclose maximum space with a minimum of building materials. Fuller never finished college, and his views were a conflicting amalgam of science, imagination, and belief. He once said that computers would save the world from ideologically dogmatic politicians, but he also dismissed the science of evolution, arguing that we were deposited on the earth in our current state from somewhere else in the universe.

He wasn't enamored with the natural world, predicting that one day humans would live in domed cities with controlled climates. While many of Fuller's views must have struck George as ridiculous, the inventor approached problems by discounting convention, and, like George, he had widely varied interests. Fuller believed that with technology, humanity could find solutions to the most difficult problems, such as hunger, poverty, and war.[2] "Buckminster

Fuller was one of the most fascinating men I've ever met," George said later. "He had a tremendous mind."[3] Fuller gave a three-day lecture—even when he taught college classes, he often spoke for eight hours straight—and the topic was one of his favorites: "spaceship earth." Fuller argued that the earth was overpopulated and that as humanity grew, resources would become increasingly scarce. "The significance of growth issues first hit me at that time," George said. He summed up the concern by saying that if the world was already struggling with 5.5 billion people, how would it handle twice that number?[4]

With his large family, George had certainly done his part in contributing to overpopulation. But he began to see that on a global scale, population growth was the root of many problems related to economics, social structures, and the environment, especially in poor communities. If these issues weren't addressed, the quality of life would deteriorate, even for developed countries. This downward spiral could threaten humanity's survival. George's time with Fuller, both listening to lectures and meeting with him in small discussion groups, influenced the oilman's emerging views on environmentalism. While direct ecological threats were a concern, George saw that they were part of a larger problem—a degradation of the natural systems that sustain all life on earth, humans included. Sustainability, George would say later, involves much more than environmental issues. Like Fuller, George believed that technology played an important role in reducing the impact on the planet of a growing population.

During his trips to Aspen, George also met the economist E. F. Schumacher, who advocated earth- and user-friendly technology that corresponded to the scale of communal life. Schumacher's rational approach to economics and his view that people should matter most in economic theory appealed to George's own pragmatism. (Schumacher also predicted the rise of OPEC, which may have caught George's attention as well.) Listening to Schumacher and others in Aspen rounded out his views. He became more convinced than ever that government and business should work together to develop sustainable societies.[5] At the same time, he worried that politicians and business leaders were too shortsighted.[6]

While he was fretting about problems facing the larger world, George tested his own theories about sustainability at home. He and Cynthia had moved from the North MacGregor house to Tynewood, a neighborhood on the west side of Houston. Although the area was considered one of the city's nicer residential pockets, George quickly noted the lack of parks. Memorial Park, the city's largest, was ten miles away, and residents had to drive four or five miles to the nearest grocery store.[7] The homes were attractive, but the developers didn't consider livability—there were too few trees and amenities. George realized that comfortable houses alone weren't enough. People wanted to live in settings that felt natural and provided a sense of community. As he saw it, too many developers in Houston were just trying to sell homes to keep up with surging demand.

The Tynewood house was temporary. Almost from the time they moved in, George and Cynthia began planning a new home on Wickwood Drive, a cul-de-sac just across Memorial Drive from Tynewood. George became obsessed with the design of the Wickwood house. He wanted it built from stone and rough-finished cedar. Cynthia put together a 247-page color-coded list of requirements, specifying everything from the linear feet of clothing storage to a prohibition against live-in servants.[8] They toured Taliesin West, the winter home of Frank Lloyd Wright in Scottsdale, Arizona. They loved Wright's modernistic style and the way the design took inspiration from the surrounding landscape.[9] To design their dream home, George and Cynthia hired architect Karl Kamrath, who was not only a friend of Wright's but also an avid tennis player. He frequently joined George at the River Oaks Country Club in Houston. Although George wasn't a member, he often played at the club with a regular group that included James A. Baker Jr., a prominent Houston attorney and father of the future secretary of state.[10]

Kamrath emulated Wright's style and his use of organic architecture, which was based on a harmony between human-made structures and the natural world.[11] The Wickwood house was 12,500 square feet and stretched some 300 feet from end to end, roughly the length of a football field. It had a swimming pool and a sunken tennis court that was hidden from the street. Four hundred fifty tons of boulders and limestone were hauled from the Pedernales River in the Texas Hill Country west of Austin to use for retaining walls and terraces. The roof was made from pebbles of crushed limestone. "It was such an artfully crafted piece of work," said Todd Mitchell, who, as George and Cynthia's second-youngest child, spent more time growing up in the house than most of the children. The house was so stunning that *Fortune* magazine featured it in a six-page spread in July 1966.

The Mitchells moved there in 1964, before construction was finished. The limestone roof design was a failure, leaking constantly.[12] But the house had all the necessities for a large family—ten bathrooms, two dishwashers, two ovens, and two washers and dryers. The children had their own wing and stayed three to a room because the two oldest daughters were in their early twenties, and several of the older children would be leaving home in a few years. George and Cynthia paid $700,000 for the home, which at the time was a princely sum—equal to about $5.5 million today.[13]

The house fronted Buffalo Bayou, a slow-moving river that undulates fifty-two miles through the heart of Houston before emptying into Galveston Bay.[14] Until the Ship Channel was dredged, the bayou was the city's most vital waterway. In fact, the city was founded on its banks about twelve miles east of the Wickwood house. By the 1960s, however, many Houstonians saw the city's bayous as a threat that needed taming. During heavy rains, the bayous often spilled over their banks, inundating the city because of its flat terrain. After a particularly devastating series of rains in the 1930s, George's old army unit, the Corps of Engineers, proposed a plan to control flooding by removing bayou vegetation, deepening the channels, and paving the banks with concrete. Just as George had decried developers' clear-cutting in the area, he didn't want to see the natural waterway turned into a giant storm drain, as the city had already done with several other bayous. He joined with like-minded homeowners and formed the Bayou Preservation Association, which believed natural trees and vegetation were the best form of flood control and a barrier against erosion. Soon after the group was formed, it was joined by Terry Tarlton Hershey, a relative newcomer to Houston who was passionate about environmental issues. Hershey was a firebrand who became a valuable ally for George. She had the drive to keep the campaign going and an understanding of the environmental issues, while George had the money to finance the campaign and the political clout to open doors with key lawmakers, including his newly elected congressman, George H. W. Bush. At the urging of George and his group, Bush held subcommittee hearings that led to the Corps rescinding its pavement plan.[15]

The bayou fight only strengthened George's belief that development could coexist successfully with nature if only politicians and businesspeople were willing to take the time—and perhaps less profit—to do it right. His opposition to channelization, though, further distanced him from Houston business leaders, most of whom backed the flood control plans. Many of the city's prominent businesspeople at the time were oil and gas executives who didn't understand one of their own siding with environmentalists. "He was willing to stand up against the whole Houston community and say 'protect this bayou,'" said Jim Blackburn, an environmental attorney who worked on several Mitchell projects. "That's not a position that many in the oil and gas business would have taken."

In many ways, George was a traditional businessman, running a company and amassing wealth. But he didn't fit in with most other executives. While George was ambitious, he wasn't brash or overconfident. He also didn't engage in traditional business pastimes like golf. While he still loved hunting and fishing, those were activities he shared with family, old friends from Galveston, or close associates within the company, rather than other executives. He never joined the Petroleum Club, the quintessential social hub for the city's energy elite.

In 1964, he built a new summer house for the family in Galveston, in his Pirates Beach development on the western portion of island. His house wasn't like any of the others. He didn't want to force a well-manicured lawn to subsist on the barrier island sand. Instead, he favored natural surroundings. "Dad was an absolute maverick among the guys who came home from World War II and mowed their lawns on weekends," George's oldest son, Scott, said. "He went down to Pirates Beach and built a home surrounded by native dunes and wildflowers. His friends told him he'd get rattlesnakes. He was the only one doing it."[16] George didn't care about rattlesnakes, and increasingly, he didn't care much about conventional real estate thinking, either.

◈

A few years after meeting Fuller in Aspen, George joined the Young Presidents' Organization, an educational and networking group for business leaders who had achieved success before age forty-five.[17] It was George's kind of group, because the members didn't just talk about business; they studied important social issues and discussed the role of business in addressing them. Many of its members were particularly concerned with the unrest from the Vietnam War and the race riots ravaging America's cities.

YPO gave George, who was in his mid-forties, deeper insight into some of the biggest social problems gripping American urban areas and encouraged him to think about solving them on a grand scale. In the oil business, he had become a voice for the independents, and he had gotten involved in shaping energy policy. YPO introduced him to a wider circle of people and expanded his understanding of the potential for business to be a leader on social issues. Instead of worrying about the next quarter's financial results, business leaders had the power to shape the world in areas beyond their expertise, such as education, poverty, crime, transportation, and globalization. It was a realization that would redirect the course of his life. He began to see the focus of business as narrow and self-serving. Too many executives worried too much about short-term financial gain rather than long-term solutions. "Corporate America has the resources, but 90 out of 100 of my counterparts could care less," he said. "Most never even expose themselves to major problems."[18] The realization would underpin a philosophy that would define his pursuit of both fracking and sustainable development.[19]

YPO was hands on, organizing trips to study these issues up close. In 1965, George visited Harlem and met young African Americans who had Harvard degrees but couldn't get loans to start a business. The next year, he went with a YPO group to tour the impoverished Watts neighborhood of Los Angeles, the scene of the largest and costliest uprisings of the Civil Rights era. Sparked by a white highway patrol officer's arrest of a young African American driver, the riots unleashed pent-up frustration from years of high unemployment, substandard housing, and bad schools. The rioting raged for six days and caused $40 million worth of property damage.[20]

After seeing Watts firsthand and recognizing the social failures that contributed to it, George decided to do something about it. "I made the decision at that time: we can do a better job in developing our cities. What's happened around the country, Philadelphia, Washington, Baltimore and all of them, the affluent [citizens] would leave the city and leave the disadvantaged to try to manage the city, which they couldn't do. We could do a better job."[21]

CHAPTER 11

Flying by the Seat of His Pants

Geologist Howard Kiatta was worried. Had he made a terrible mistake leaving a good job at Texaco in Louisiana to join George Mitchell & Associates? Though he'd been born and raised in Houston, Kiatta had never heard of George Mitchell before he took the job. Still, one of his former colleagues, Tom Purcell, was already working at GM&A and spoke highly of George. Everyone he asked had told him that George was the smartest oilman in town. Kiatta liked the idea of joining a dynamic, growing independent oil company. He didn't want to spend his entire career at Texaco, as his father had. But soon after he signed on in 1967, Kiatta learned that George was known for something other than his intelligence.

Kiatta called a company in New Orleans to order a geological map for a review of potential drilling prospects in southern Louisiana. The man on the phone took his request and asked who he worked for. "George Mitchell & Associates," Kiatta said. The man told Kiatta to send a check first, and then the company would send the map. "Can't you just invoice me?" Kiatta asked. No, the man said, because George Mitchell had a reputation for not paying his bills. A few weeks later, Jade Oil filed for bankruptcy, and Kiatta wondered what sort of organization he'd landed in.[1]

Kiatta's friend Tom Purcell, another geologist who'd joined the company a few years earlier, said GM&A was quick to invoice but slow to pay, in part because George never used a standard budgeting process. All prospective drilling deals went straight to him for evaluation. "He'd either say 'yes, but don't pay over a certain amount,' or 'I'll think about it,' which I took to mean he didn't have the money," Purcell said.

Purcell witnessed his boss's freewheeling approach to finances firsthand. In the late 1960s, the company was planning to submit bids to Texas regulators for two or three tracts that Purcell thought looked promising. The bids were due at the Texas Railroad Commission, the state's oil and gas regulator in Austin, the following Monday. George planned to be out of town, so on the preceding Friday he wrote two checks totaling $27,000 and put them in his desk drawer with instructions that they be sent to Austin along with the bids on Monday. GM&A never submitted the bid. When Purcell came into the office Monday morning, George was already there. He had canceled his weekend trip. "He said, 'I've *got* to have those checks,'" Purcell recalled. "He had spent the money on something else over the weekend." Purcell never found out on what.[2]

By this time, GM&A was a good-sized company. It had $250 million in assets and a tremendous amount of debt, yet it had no formal process for budgeting or evaluating acquisitions. "Everybody was flying by the seat of their pants," said Jim McAlister, whom George hired to help bring more structure to the finances. "George was comfortable flying by the seat of his pants. He carries files in his head that most people don't remember are in their filing cabinets. The company was operating as a projection of George's personality. Well, we had to come out of that. We were growing, we were getting bigger, we had to have a budget and better-defined investment criteria."[3]

George may have lived in a custom home worthy of a six-page spread in a national magazine, but his views about money hadn't changed much since he was a poor kid hustling unlucky fishermen on Bettison's pier. He liked earning money and he hated letting go of it, regardless of the amount. He faithfully left the office at 3:30 p.m. to play tennis. He had a standing agreement with his tennis buddies: the last one to the locker room bought a can of balls and a round of beer after the game.[4] "He didn't want to spend $2 on a can of balls," Kiatta recalled.[5]

Before he joined GM&A, Purcell worked for Tenneco, first in Houston and then in Corpus Christi. He and his wife wanted to move back to Houston, so over Thanksgiving in 1964 he interviewed with GM&A. George asked him to start the next week. Purcell agreed but mentioned that Tenneco typically paid him a Christmas bonus. He was hoping that George would simply cover the bonus, but instead, George asked when Tenneco typically paid it. The Friday before Christmas, Purcell said. "Why don't you just stay down there and come up after the first of the year?" George replied.[6]

George's tightfisted reputation dated to the early days of the Boonsville Bend, when the company was still operating on a shoestring and he and Johnny would rush to the bank with the latest royalty checks to cover the cost of the next well. Soon after the firm built the Bridgeport gas plant, George hired a consultant, a friend of his brother Christie, to study the feasibility of selling ethane, butane, and other gas products to chemical companies along the Houston Ship Channel. The consultant became a lifelong friend, but despite sending multiple invoices, he never collected his payment from George. When the consultant told a mutual acquaintance that he had done work for George and was trying to get paid, the man shook his head and said, "What were you thinking?"

At other companies, under another executive, such lackadaisical accounting might have unnerved most employees, but George's easygoing manner generated intense loyalty among his employees. He didn't rely on fear and intimidation. Instead, he was approachable and took a personal interest in their lives. He was the rare executive who inspired many of those who worked with him, creating a workforce that trusted him deeply.[7]

Bryan Smothers started in the mail room in 1965 right out of high school and later got promoted to the geological drafting department. He took night classes to learn drafting, but in 1967, he himself was drafted and wound up serving in the 101st Airborne in Vietnam. George wrote him regularly, giving him updates about the company and occasionally even sending fifty-dollar money orders, telling Smothers to use it on his next leave in Saigon. (George, who'd never served overseas himself, didn't realize that Smothers never left the jungle during his tour of duty.) When Smothers returned from the war, he found that George had kept a job for him. Smothers had picked up photography as a hobby in the army, and George eventually made him the company photographer for The Woodlands.[8]

While George was quiet and even shy in public, in the familiar confines of the office he was often more animated, especially with Budd Clark. The native New Yorker could be foulmouthed and bellicose. He and George respected each other, but Clark was a pragmatic number cruncher and George was a visionary and dreamer who rarely dwelled on the details. Clark's job was to bring some discipline to George's seat-of-the-pants style, which led to frequent clashes—often loud ones. As they made their rounds each morning, getting the latest information on various drilling programs, managers always knew George and Clark were coming. "They would just air it out in the hallway," Kiatta said. "That was how they settled differences. They would just yell it out." After the staid culture of Texaco, Kiatta wasn't used to the freewheeling style he encountered on the twelfth floor of the Houston Club Building.[9] "I thought they were going to kill each other, but it turned out that's just the way they would speak to each other," he said. In time, he came to realize that George and Clark were actually quite close.[10]

The two disagreed on politics and current events. George, for example, opposed the Vietnam War; Clark supported it.[11] "I don't think any of it was vicious," said Joyce Gay, who was George's secretary and later became a vice president who handled everything from public relations to the selection of office artwork. "I really and truly think people were enjoying themselves when they were screaming, and it was just a management style." The management-by-yelling culture took some getting used to. Others in the company adopted the high-volume management style simply as a way of being heard. "They'd interrupt you in mid-sentence, and if you let that stop you, you'd never get to tell them what you wanted to tell them," Gay said. "You pretty much had to jump right in." As a result, management meetings often devolved into shouting matches. During one such high-volume gathering with an outside attorney, the lawyer came out of George's office, shut the door behind him, and sat down near Gay's desk. He said he wasn't going back in until everyone stopped yelling. "It was a good five or ten minutes before the rest of them realized he wasn't there," Gay recalled.[12]

George's demeanor with the rank and file was usually quite different. He projected an easygoing manner that most of his employees appreciated. He disliked reprimanding people publicly, and even when he had to do so, he tried not to embarrass them. His employees knew that he placed a lot of faith and responsibility in them, and he could be demanding. In discussing the latest steps on a project, he might simply tell an employee to "take care of it." "There was a level of intentionality and determination with George, and he wouldn't have succeeded without it," environmental attorney Jim Blackburn said. "But there was a hard side to him."

In the 1970s, George was supposed to drive a politician from Houston to Galveston in his Cadillac. The meeting was scheduled for a Monday, but over the weekend, George went fishing and threw a stringer of fish in the back of the car. By Monday morning, the stench was unbearable. He told his assistant to get the car cleaned, but after several attempts, the smell only grew worse. The dealership offered to give George a loaner vehicle for the trip, but he was insistent. He wanted his car. Finally, the dealership replaced all the carpet and delivered the vehicle just minutes before the politician arrived.[13]

Most mornings, George went to the coffee shop on the building's ground floor, but his schedule was irregular. Employees who wanted coffee ran the risk of being spotted and peppered with questions. "It didn't matter which door you came through—there were two doors—he was going to pop you with a question because he knew what everybody did, what the status was, and he always had questions," said Earl Higgins, who joined the company just after Christmas 1967 and would later become vice president of property management for The Woodlands.[14]

Drilling for oil and gas is a nerve-racking business, and despite all the analysis and data that can be gathered on a prospective well, there's still an element of guesswork. George understood this, and while he demanded a lot from his employees, he never got angry if they drilled a dry hole. He would simply tell his engineers that they would do better next time. Once, Kiatta walked into George's office, stood in front of his desk, and told George the latest well was a failure because they had drilled on the wrong side of an underground fault. George replied, "I know where the fault is, it's on the other side of this desk." Then he laughed and went back to work.[15] The way George looked at it, even dry holes taught the company something about drilling in a particular area, and the money the company lost was simply the cost of gaining knowledge. As a result, he was willing to invest in failure, at least up to a point. And since George was always looking years down the road, money spent in the short term was simply an investment that might not pay off for decades.

When Purcell joined the company, he had been studying a prospect in Polk County, Texas, northeast of Houston. The area wasn't known as a big oil producer, and George had little interest in it. In one of the few budget meetings the company had during Purcell's tenure, they were reviewing different potential projects, and Purcell again suggested his Polk County idea. George decided they should table it. Purcell knew that was George's way of killing the deal, and he protested, telling George it was the best deal they had in front of them. He even offered to find another partner to shoulder half the costs—provided George paid invoices promptly.

The first well they drilled blew out, sending a huge plume of flame shooting skyward. The accident made logging the well impossible, and some of the partners Purcell had brought in wanted to simply plug it and forget the whole project. Despite the problems, gas was coming out of it, which convinced George that Purcell had been right. He authorized drilling a second well, which wound up producing almost twenty billion cubic feet of gas.[16]

◈

George had become one of the largest and most successful wildcatters in Houston, and with the size and success of the company came a degree of credibility. He and Johnny had been the voice of the independents, but most of those were small, private companies that couldn't compete with the megaphone of the major oil companies. George wasn't really part of the oil establishment, even in Houston. It went beyond his preference for eating at the Houston Club rather than the Petroleum Club.[17] He and many other executives at Mitchell Energy had little involvement with professional organizations. George simply didn't have time, and what spare time he did have he devoted to fishing, tennis, or his interest in the emerging field of sustainable development.[18]

Yet with the growing concern about imports among the independents, his company's sheer size gave George a platform. He wasn't the public agitator that Johnny had been, but increasingly, people inside and outside the industry wanted to know his opinion about energy policy. George's stature in the business community extended to the political arena as well. Texas governor Preston Smith came to George's Cape Royale development on Lake Livingston for a catfish cookoff.[19] Senators and other dignitaries began visiting the offices regularly, and the Houston Club digs were outdated and overcrowded.[20] "We were packed in like sardines, desktop to desktop," Earl Higgins recalled. George wanted a new headquarters that reflected the company's growing prominence in the energy business.

In 1971, GM&A moved into One Shell Plaza, a fifty-story skyscraper in downtown Houston, just a few blocks from the Houston Club Building. As the name implies, the building served as the headquarters for Royal Dutch/Shell's US operations, which had come to Houston from Manhattan a year earlier. It was also an important symbol of Houston's emerging role as the world's energy capital. The tallest building in Texas at the time, the skyscraper had an exterior of white travertine marble mined from the same region as the stone in the Roman Colosseum. Its gleaming facade stood as a beacon for businesses drawn to the modernity of Space City.

The plans for the building had begun six years earlier, when oil prices were in a slump and Shell found Houston's low cost of living attractive. The company transferred about 1,400 workers from New York and hired hundreds more locally, but even so, it wasn't enough to fill the entire edifice.[21]

For George, moving into One Shell showed that his company belonged among Houston's most prominent energy firms. After all the political battles between the majors and the independents, they would now be under same roof. GM&A had two full floors—the thirty-eighth and thirty-ninth, which was where George and other top executives had their offices—and parts of four more. At an opening gala, employees and their families and friends wore buttons that said "Hello, I'm with George." George wore one that said "Hello, I'm George."[22]

He wanted to infuse the office with contemporary design that reflected the modern building. He hired an interior decorator and had his office painted his favorite color, orange, and furnished it with contemporary seating and minimalistic wall hangings.[23] (His love of orange was at odds with his ties to Texas A&M. Orange was the color of A&M's archrival, the University of Texas.)

Most of the employees went from having individual offices to working in an open "bullpen," a concept that was coming into vogue at many companies at the time. Tom Purcell had never seen an office like that.[24] Howard Kiatta hated it. "I just couldn't stand the noise and the distractions and the lack of professional feeling that came with working in a cubbyhole where we didn't have any walls," he said. Senior executives, though, still had corner offices. George's was in the southeast corner of the thirty-ninth floor, and Kiatta managed to get the "cubbyhole" next to it, which gave him a lot of contact with the company's namesake. After listening to Kiatta complain about the noise and the office situation for months, George offered him an office two floors below. Kiatta thought about it for a few minutes and decided against it. "I didn't like the office arrangement, but I sure did like being next to George," he said. "I wouldn't have traded it for anything."[25]

Every morning, George would step off the elevator and walk down a long hallway toward his office. If he encountered any employees in the hall, he'd stop them and either ask them a question about a project they were working on or tell them something he wanted them to do. As he walked into his office, he would relay to his assistant Linda Bomke (Joyce Gay had been promoted by then) the names of those he wanted to play tennis with that afternoon. She had a list of his tennis buddies and worked her way down the names until she had assembled a foursome. George played every day and most weekends as well, although his weekend matches were usually in Galveston.

George would look over a list Bomke had made of phone calls that had come in since he had left the previous afternoon for his tennis game. The document had a name, a phone number, and a one- or two-sentence synopsis of what the person wanted. Below that was a tally of people that *he* wanted to talk to that day and their numbers. After glancing at both, he would head down to engineering on the thirty-eighth floor, making the rounds from one department to another, getting updates on every project and every well. His last stop was geology, usually Kiatta's cubicle since it was next to George's office. The entire circuit of both floors typically took him twenty minutes.

Then he would return to his office and start making calls. George loved the telephone, and he had no compunction about calling employees at night or on weekends if he thought of something he wanted to tell them. His phone calls during the day, as well as his meetings, spanned numerous projects. "It was nothing for him to have 12 or 13 projects going on at the same time," Bomke said. He expected all his managers to understand that no one should work exclusively on one project and stay with it until it was finished. They worked on several simultaneously, because that was how George's mind worked. He could do many different things at once, and although he remembered details of every project, he didn't micro-manage. He knew the outcome he wanted, but he allowed his managers to decide on their own approach to a project and rarely bothered with the minutiae of how it was done.

Midmorning, he would take his fifteen-minute break for a cup of hot tea in the lobby coffee shop, usually bringing a randomly selected group of employees with him and peppering them with questions about different projects. He loved mingling with the workers and talking to them about what they were doing. He took a similar break in the afternoon.

If he had a business lunch, he scheduled it at the Houston Club. Otherwise, he rounded up people in the office and took them to a nearby restaurant, usually Steve's Barbecue, a rundown but well-known establishment three blocks away. George could get through the serving line at Steve's, eat lunch, and be back in the office in less than thirty-five minutes.[26]

He never carried much cash. His employees would file through the line at Steve's ahead of him, and when he got to the cashier, he would inevitably ask if anyone had money. Kiatta lost track of how many times he had to pick up the tab for lunch. "He was one of the richest guys in town and he's borrowing money from me to buy lunch," Kiatta said with a tone of disbelief, forty years later. Unlike in his dealings with some of his vendors, though, George paid his employees back quickly.[27]

His workday ended promptly at 3:30, so that he would have time to make it to his tennis game by 4:00 p.m. (In the winter, when daylight savings time ended, the schedule shifted up by half an hour.) If he was in an afternoon meeting, Bomke would stand in the doorway and tap her watch at 3:30, a signal to George that it was time to leave. All the employees knew the sign, and they knew to discuss any pressing matters early in the meeting. Nothing would keep George from making his tennis match on time because of the standing deal that the last player bought the balls. No one wanted to be the one who made George late for his match. Company lore was rife with tales of those who failed to heed the schedule or wound up following George to the parking garage, finishing a conversation or a meeting as he walked briskly to his car. One winter day, Tom Purcell wandered into George's office a little after 3:00 p.m. He'd been wrestling with a big decision, and he needed to discuss it with George. He walked in just as George was walking out. Purcell tried to buttonhole his boss, asking for a minute to talk. George said he was trying to leave, and whatever Purcell wanted to discuss could wait until the next day. "George," Purcell said, "I'm trying to tell you that I'm going to resign from the firm." George looked at his watch and said, "We'll talk about that tomorrow. I'm playing tennis."[28]

Purcell did eventually leave the company to strike out on his own, but before he left, George stopped by Purcell's desk one afternoon. He'd noticed that Purcell was using aerial photos to study drilling prospects. George wondered how he might get some for some land he'd been looking at for a different project. Purcell said all he needed was the coordinates of the area George wanted to see. George gave him the description of some of the land he'd bought in Montgomery County, north of Houston. Purcell thought the request was odd. No one had even suggested drilling up there. It was mostly timberland.

When it arrived, the aerial photo was about three feet by four feet. George hung the picture at the end of the hallway near his office. He would stare at the photo, sometimes from several feet away, and other times would look at it closely with a magnifying glass. "Every now and then, I would see him standing down there, and I'd walk down and talk to him, but it was obvious he didn't want me asking any questions about what he was looking for," Purcell recalled.[29]

George himself may not have known what he was looking for, but he gazed at the picture as if he were looking far into the future.

CHAPTER 12

Lafitte's Gold

By the mid-1960s, more companies were actively searching for oil and gas outside the United States, but George remained skeptical of international expansion. In addition to the Boonsville Bend, GM&A had developed a major presence on the Gulf Coast and in South Texas, and it added acreage in Ohio, the Rocky Mountains, and California. In the late 1950s, it made some investments in Mexico. Oil in the ground tends to collect near salt domes, upward-pressing deposits of salt or halite that cut into the geologic strata above it. Other minerals often collect near the domes as well, and the company found a large sulfur deposit in Mexico. No sooner had it formed a subsidiary, Sulfurex, to develop the prospect than the Mexican government seized the assets.

George had a similar experience in Canada several years later. While drilling for oil and gas, the company found a large deposit of potash, a form of mineralized potassium that's used for fertilizer. Just as in Mexico, it lost the assets when the government of Saskatchewan nationalized the potash industry in 1967.[1] "He would talk about how bad he was screwed by Sulfurex and the potash deal," his second-youngest son, Todd, said. George hated to see a big deal fall short of its potential, and the loss of both projects frustrated him.

In the 1980s, Gene Van Dyke, a Houston wildcatter who lived near the Mitchells, convinced George to invest in a drilling program Van Dyke was putting together in the North Sea.[2] Van Dyke made a big discovery, but George wasn't used to the cost of offshore projects, and the North Sea was a harsh and difficult environment for drilling. Even today, it has some of the most expensive offshore wells in the world.

When Van Dyke told George the price for installing a production platform, George said, "Hell, we can't afford to make a discovery" and sold his interest.[3]

"Sulfurex, potash, the North Sea—those are the things that made my dad not interested in international work," Todd said. "In the international game, you never know what will happen legally or with the government, and if you find something, you're a long way from getting it onto the market because there's huge infrastructure and capital costs and [he thought] it's just not the game we should be playing."

Even in the United States, though, the nature of drilling projects was changing. As oil prices rose in the 1960s and 1970s, companies could afford to invest in more expensive and technologically challenging techniques. Kerr-McGee, an Oklahoma company, drilled the first true offshore well in 1947 about ten miles from the coast of Louisiana. At first, companies were reluctant to embrace the idea.[4] Offshore wells cost about five times as much as land wells, and the majors found it cheaper to import from the Middle East rather than wrestle with the expense and uncertainties of this new technology. Gradually, though, as the costs of offshore drilling declined and regulations became more defined, companies started plying the waters of the Gulf of Mexico.

In 1967, Gulf Oil completed a seismic survey of Galveston Bay that included parts of Pelican Island, which George had bought earlier in the decade. A seismic survey is like an ultrasound of the earth—sound waves are bounced off underground rock formations in the planet's crust. By analyzing the waves, geologists can identify different rock layers and other geologic clues for potential oil and gas deposits. Gulf asked George whether it could extend its survey onto his island. He agreed, on one condition: Gulf had to share the data with him. Although his company was now worth millions and he had hundreds of employees, George hadn't changed that much from the eager young engineer studying borrowed well logs in the wee hours. If he could find a way to get his hands-on data without having to pay for it himself, he would.

Gulf handed over the Pelican Island information, and although the company never pursued its own drilling program, George and some of his geologists, including Tom Purcell, pored over the survey themselves. The island didn't appear to hold any significant deposits, but they spotted an unusual tilt to the rocks beginning in the bay just east of the island and running toward the open Gulf, right under Galveston. If they were right, George's hometown was sitting atop a significant reservoir of oil and gas.

Gulf's survey hadn't included Galveston Island. The mapping offered little more than a geological hint, which wasn't enough to justify the expense of an exploratory drilling program. If George wanted to know for sure, he'd have to shoot his own seismic survey under the city. That would be a problem. To collect seismic data, something must generate the sound waves that bounce off the rock. Today, compressed air can create the necessary vibrations, but in the 1960s, that technology didn't exist. Instead, most companies detonated a small charge of dynamite. To get the results George needed, he would have to set the explosives right down Broadway, the main thoroughfare across the island, which was home to about sixty thousand people. It was so ridiculous that the only question was whether the city council would have a good laugh before they rejected the whole idea. George, though, wasn't easily deterred.

He ruminated on the problem. What if they found an alternative to dynamite? A decade earlier, Conoco had developed a technology called Vibroseis, in which a steel plate was mounted on the undercarriage of a heavy truck. Once the truck was in place, the plate was lowered until it contacted the ground, and motors then caused the plate to vibrate, creating sound waves through the ground that served the same purpose as dynamite. Purcell agreed the idea might work. It would be loud, but far less disruptive than dynamiting the center of town.

Perhaps because of his stature as Galveston's wealthiest native son, George convinced the city to allow him to proceed. However, he could use the Vibroseis trucks only at night. George also agreed to pay for police to divert traffic away from the area, because vibrations from passing cars could disrupt the seismic data. The trucks did their work at 2:00 or 3:00 a.m., and they produced data good enough to confirm the trend that the Gulf survey had hinted at. George now knew there was oil and gas under Galveston.

Most other companies still would have walked away. Drilling in the center of a midsized city was daunting. Drilling rigs are loud, they run at all hours, and they can unleash noxious smells from the bowels of the earth. Residents would complain. No city would agree to such disruptions, and even if one did, the restrictions would be so stringent that it wouldn't be worth the trouble. But this was George, and it was Galveston.

Since Sam Maceo's death in the early 1950s, the island had been on an economic slide. The state's attorney general, backed by a squad of elite Texas Rangers, raided the island's gambling dens soon after Maceo died. Grand juries indicted more than one hundred people, including members of the Maceo family. The Rangers rounded up slot machines and roulette wheels, loaded them on a barge, and, for the benefit of the press, dumped them into Galveston Bay. Another monster storm, Hurricane Carla—four hundred miles across with wind speeds that reached 173 miles per hour—smashed into the island in 1961.[5] The storm touched off a series of tornadoes in its path, which combined with the storm surge that flooded the island and added to the devastation. Rebuilding efforts lacked the commitment that had prevailed after the 1900 storm. The Houston Ship Channel had long ago siphoned off the port business, and the night spots of the Maceo years had fallen into decline. Beachgoers flocked to new hot spots like South Padre Island or the Florida Panhandle. The Strand, once known as the Wall Street of the Southwest, became run-down and deserted except for the occasional drunk or hobo. Hendley Row, the most prominent stretch of commercial buildings in the Strand, became derisively known as the "Wino Hilton." Real estate prices fell, and property owners grew increasingly desperate to sell. The city was fading away.[6]

George knew that his discovery, if he could get it into production, could give islanders and their city a much-needed economic boost. He told his engineers to begin leasing mineral rights. In Texas, state law separates the ownership of mineral rights from surface rights. That means that in many newer cities, homeowners may control who drives on their property, for example, but someone else holds the rights to oil or gas underneath their land. The property owners have no say in drilling if the mineral owner wants to drill. Nor does a homeowner reap economic benefits if oil is discovered. Galveston, however, was different. Because it was one of the state's oldest cities, most of the property owners also held mineral rights, meaning that if he wanted to drill on the island, George would have to lease rights from thousands of individual homeowners—as many as six separate leases per acre. And George planned to lease at least 2,500 acres. His land men estimated that acquiring all the necessary leases would take a year and a half and cost about $350,000. George suggested they start small and keep the project quiet. With so many leases needed, he didn't want word of the drilling to get out for fear other companies would start snapping up mineral rights on the island.

George's team started with fifteen acres in the Southern Pacific rail yard near the Strand, on the bay side of the island. It wasn't much, but at least they had a foothold. Like the rest of the island, the rail yard was a shadow of its former self. While many of the tracks were largely unused, a few switch engines still rolled around, moving freight cars from one track to the other and hooking up trains taking goods from Galveston's port to the mainland. George drilled that first well between the tracks. Before they had a chance to acquire more leases, he walked into Purcell's office and said, "Hell, it's probably going to be a dry hole. Let's drill it anyway." Purcell was shocked. If the well was successful, everyone would know—well results had to be reported to the Railroad Commission—and the leasing costs for the rest of the area would skyrocket. "But there was no way he was not going to drill that prospect," Purcell said. As George was leaving the office, he stopped and turned back to Purcell. "We're going to tight hole it," he said. "Don't tell anybody." A "tight hole" meant that almost no one in the company—and certainly no one outside it—would know the well was being drilled.

Purcell didn't understand his boss's reckless determination. George had started in the business as a wildcatter, but he had never been the sort of freewheeling gambler that many other independent oilmen were. His decisions to drill were usually based on careful analysis. Nevertheless, Purcell did as he was told. He and George were the only people who knew the project was going forward in such an unorthodox manner. Even the head land man, Jack Yovanovich, wasn't told.

They drilled it in the middle of the rail yard and no one in the industry seemed to notice. The night they got the log back, George did something he rarely did—he came out to the well site. Even much later, in the 1990s, as his engineers perfected fracking and the future of the company hung in the balance, George didn't visit wells. But this was different. This was Galveston. This was home.

He and Purcell went into the trailer and asked the other workers to leave. They rolled out the log on the table and looked at the squiggles of red and blue lines that reported the drill bit's journey like some geological travelogue.[7] The lines showed the depth, porosity of the rock, and other characteristics. Purcell studied the log and realized they had a winner. He looked across the table and saw tears running down George's cheeks. "The hometown boy made good," Purcell said, tearing up himself at the memory almost half a century later.

George was known to push his team in the face of overwhelming odds, but from the beginning, this project was special. He had left Galveston years earlier to make his fortune in Houston, but Galveston had never left him. He knew as well as anybody the struggles of life on the island. He had grown up there, motherless, working from the time he was a boy to help provide for his family and to scrape together a little money for himself. He had relied on the generosity of Sam Maceo to get through college. And he knew that despite all the hardships, he was one of the lucky ones. Galveston was an island divided by economics. Wealthy Houstonians bought fancy homes on the west-end beaches, but in town, behind the safety of the Seawall, people still toiled and scraped just as Mike Mitchell had. George wasn't drilling this well just for himself. Over the next year and a half, he would offer thousands of homeowners a chance to benefit from the natural gas field he had found. It was an economic shot in the arm for the whole city, and while it didn't turn any paupers into millionaires, it offered something many island residents hadn't had in their lives—a lucky break.

"Nobody but George would have tackled that," Purcell said. "Nobody would have drilled it with that land situation. This was the one exception to his risk aversion—for Galveston." As they looked down at the well log, and the potential of the find sank in, George made a declaration. Just as he didn't typically visit well sites, he rarely got involved in naming discoveries, which tended to take their appellations from geological characteristics, as "Boonsville Bend" had. But this time, he picked the name—Lafitte's Gold. It referred to the famous nineteenth-century Gulf Coast pirate Jean Lafitte, who used Galveston as a base of operations from just after the War of 1812 until 1821. Lafitte was a mythical figure in the island's history.[8] Two centuries after his death, rumors persisted that he had hidden treasure somewhere near the city. The parallel between pirate booty and black gold was too strong to resist.

George's order for a tight hole remained in place. He and Purcell locked away the well log in a safety deposit box that required George's signature to open. They disassembled the drilling rig, plugged the well, and installed a metal plate over the hole. George instructed one of the crew to plant some wildflowers at the site so anyone walking by would be less likely to see it. Then, the leasing program began. George hired a crew of people to scour the island neighborhoods. Land brokers walked the streets with soft leather satchels over one shoulder containing the leasing papers and with a wad of fifty-dollar bills in hand—one for each person who signed a lease.[9] They eventually collected leases from more than five thousand residents—basically everyone who lived in the heart of town, between Thirty-First and Fifty-Ninth Streets.[10]

As the last of the leases came in, George got an urgent message from the company's lawyer. When he got to the office, the lawyer pulled out a letter from the Railroad Commission. The regulator said it had never received a completion report—which operators must file on every drilling project—for the well in the rail yard. The completion reports are public record, and other oil companies watched when a competitor drilled in a new area. George didn't want potential rivals knowing what he was up to. Apparently, his "tight hole" directive went farther than just planting some flowers over the wellhead. The lawyer stressed that the letter was serious. The commission had the authority to fine the company $100 a day from the time of completion until the report was filed. Perhaps George did the math in his head and realized the fine would be less than $55,000—a pittance compared to what he was spending on leases and drilling costs. His response was dismissive. “Forget it,” he said as he walked out of the office.

Even though he had the leases in hand, George knew he couldn’t just set up a drilling rig in the middle of the street. And unlike the seismic tests, this work couldn’t be done in the dead of night. Instead, he bought an old cotton warehouse on the north side of Broadway, had the roof taken off, and put the rig inside. From there, the company drilled directional wells under the city. Directional wells, sometimes called slant-hole drilling, are exactly what they sound like: the well is drilled down vertically, and then at a certain depth, it’s angled or slanted to the side. From the single well site, the company drilled more than a dozen slant holes in different directions. Most people in Galveston had no idea what was happening 7,000 to 9,500 feet beneath their homes.[11]

Lafitte’s Gold wound up being one of the biggest discoveries in the company’s history. It paled compared with the Boonsville Bend, but by the year 2000 Lafitte’s Gold had produced more than 1.6 billion barrels of oil and 89 billion cubic feet of natural gas. As for the individual residents of Galveston, their payouts were relatively modest. When the company was still signing leases, it ran ads in the Galveston paper showing a drilling rig with oil drops raining down from it emblazoned with dollar signs. “Oil on the Island!!” it blared. “It could be worth real money to you.” “Real money” meant “$10 to $15 a month for 15 years.”[12] In today’s dollars, that would be between $66 and $99.

The biggest contribution was to Galveston’s overall economy. Between 1972, when the first well from Lafitte’s Gold came in, and 1989, George’s projects paid out $11 million in royalties. Sixteen percent went to the city, and the rest went to residents who’d leased their mineral rights. George employed more than one thousand people on the island and paid more than $1 million a year in local taxes. That was significant for a community with relatively few middle-income jobs and a tax base supported mostly by mom-and-pop businesses. (The biggest employers in town—the wharves and a University of Texas medical school—were tax exempt.) George’s drilling program made him one of the city’s biggest benefactors and emboldened his investment interests in Galveston.[13]

More important, Lafitte's Gold had a direct effect on future projects. Not only did George and his engineers overcome the engineering, political, and public perception challenges; he also demonstrated that wells could be drilled in a heavily populated area without harming the environment or disrupting residents' lives. Unfortunately, the lessons of Lafitte's Gold were largely lost on an industry that, less than four decades later, would face outcry and even drilling bans from other communities that found their homes, schools, and parks in the path of fracking.

◈

If it hadn't been for George's real estate investment on Pelican Island, Lafitte's Gold might never have been discovered. Over the years, he'd been steadily buying up land on the island, including Pirates Beach, his 1,200-acre residential development on the west end where he had his weekend home.[14] Bob Smith may have made his first fortune from inside the land and his second on top of it, but George was finding the relationship between drilling and developing to be more symbiotic. He began studying the geology of the island in more detail and realized that Lafitte's Gold was just the beginning. The formations appeared to extend east from the island out into the Gulf of Mexico. Two of the most promising places to drill were on the beach and about a mile offshore, directly in front of Gaido's, Galveston's most famous seafood restaurant, where George had sold fish and cleared tables as a boy.

George bought a twenty-two-acre site on the Seawall that had been the location of Fort Crockett, an old army coastal artillery bunker left over from World War II. The gun emplacements included elevated bunkers with seven-foot-thick concrete walls. He planned to drill behind the bunker and then angle his wells into the Gulf. Tourists, he told the city council, would never know what was happening. City leaders weren't convinced. George personally lobbied business owners along the Seawall, and he bused in hundreds of his own employees from Houston to pack the council chambers and give the appearance of widespread public support. He got well control expert Red Adair—made famous in the John

Wayne movie *Hellfighters*—to testify in support of the project. But even Galveston's most illustrious native son couldn't get everything he wanted. The council rejected George's drilling plans. So, he got offshore concessions from the state and went on the water, a mile from the beach and the Seawall. The offshore wells cost double what he would have paid to drill on land—in a concession to city leaders, George agreed to shut the rigs down between April and October, prime tourist season—but the projects were still profitable.[15] The company built a special rig, one that was less visible from the shore. "It cost a lot more to do it the way we did offshore, but if you go to Galveston . . . you don't see any ugly platforms," George said. "It looks like a little buoy." He estimated his hometown earned about $2 million in taxes and another $9 million in royalties from the project, but it drew the ire of environmentalists, which frustrated George because he had gone to such lengths to minimize the environmental impact.[16]

As for Fort Crockett, George refused to sell it even after being denied the right to drill from there. It may not have worked as a drilling prospect, but the site was still prime real estate. He would hold on to it and see whether he could come up with another use for it.

CHAPTER 13

Going Public

Shaker Khayatt was an immigrant success story. Born in Egypt, he came to America in 1957 and enrolled in the Massachusetts Institute of Technology, earning a master's degree in engineering two years later. He went to Harvard Business School before taking a job on Wall Street with the boutique investment firm F. Eberstadt & Co.[1] In the late 1960s, he got a call from a loan officer he knew at Chase Manhattan Bank. The banker had a client in Texas that he thought Khayatt should visit. Sooner or later, the company would need an investment bank to take it public, the loan officer said, because it was heavily indebted and would need access to cheaper capital.

Khayatt came to Houston, met with George and Budd Clark, and began looking at the company's books. He was surprised at what he saw. Not only was there a lot of debt, but some of it—for $1,000 or less—was owed to various individuals or small companies. One of the more usual line items was mice. George had learned that cancer researchers were facing a shortage of laboratory mice because no one wanted to pay for a breeding program. He believed the research was important and that somebody should fund it, so he put up the money. "I told George: 'Please, George, no mice. I've got a lot of friends on Wall Street, a lot of credibility, but it doesn't extend to mice,'" Khayatt recalled.[2] He convinced George to make those sorts of investments privately, not through GM&A, especially if the company wanted to go public. At the time, George had not established a family company for his personal investments. He just ran everything through GM&A.

Mice aside, the cash demands for the North Texas drilling program had become too great for George to finance with private investments and debt. The company had to keep drilling to meet the terms of its contract with NGPL, and the only way it could raise enough cash was to tap the public markets. Getting GM&A ready for the scrutiny that came with a stock offering wouldn't be easy. After years of George running the company as his own and doing business with friends and family like Johnny, GM&A was a complicated web of related-party transactions. Not only did the books have mice, they had countless other deals that seemed to have little to do with the oil and gas business. George's wide-ranging interests had slowly infiltrated the company's finances. By then, GM&A also owned sizable real estate holdings in Galveston and the timberland north of Houston. Were investors buying a real estate company or an oil and gas company? That question would plague the company for the next twenty-five years.

Then there was the structure of the offering that Khayatt had put together. He had told George to sell most of the company and keep about 35 percent of the equity for himself. George refused. "I said, 'no, I'm not going to listen to these people.' Had we sold down to 35 percent, when the first downturn came, it would have been gone," George said later. The oil business was cyclical, and George had seen too many other companies struggle with financial hardship in the slumps only to get bought on the cheap when their stock prices fell. By keeping a controlling interest, he could prevent an unwanted takeover.[3] While that protected the company, it also made the deal less appealing to investors. It meant that regardless of the quality of the operations, they were betting on George himself, and outside Houston, few people knew who he was.

Typically, when a company goes public, the initial stock value, or offering price, is set based on the values of comparable companies. But George had his own idea of what his company was worth, and he told Khayatt that he thought it should fetch between $17 and $18 a share. For a publicly traded oil company, GM&A was still relatively small, and the market value was closer to $14 to $15. Walter Lubanko, a lawyer who worked on the deal and who would later join the company's board, came up with an idea. The company would package its common stock with other securities such as warrants, and debt known as convertible debentures—which could be converted to common stock—and sell it all as a single unit. The units would have the combined value of $17 a share, and George would retain control over two-thirds of the stock. The company hoped to raise about $25 million in the offering.

The timing, in late 1971, wasn't good. Investors had little appetite for midsized oil companies, and the complexity of the arrangement, combined with George's insistence on retaining control, was more than the underwriters would take on.[4] The deal had to be approved by an underwriting committee—twenty-five financial firms that would sell the stock. Twenty-four voted to reject it. Prominent Wall Street firms like Goldman Sachs were offered a chance to participate in the offering and declined. The committee sent two young representatives to Houston to tell George he needed to sell some assets and reduce his debt. The meeting ended abruptly.[5] George didn't need some young investment banker lecturing him about debt. More importantly, he hated selling assets. You never know, he used to tell his geologists and engineers, what might happen.[6]

Khayatt decided to try again with a simpler plan—stock only. This time, he chose a dual stock structure, with Class A shares controlled by George and Class B shares sold to the public. The Class A stock had more voting rights, which meant that George could get the benefit of raising money from the stock market without having to worry too much about pleasing his investors. The structure was popular, especially among family-run businesses going public in the late 1960s and early 1970s. The Class B shares would trade based on the company's financial performance, but they were still beholden to George and his decisions, mice and all.

As part of the offering, the company changed its name again, to Mitchell Energy and Development Corporation. The name reflected the company's split focus on oil and gas as well as real estate. The shares began trading on the American Stock Exchange at about 10:30 a.m. on February 26, 1972. At noon, Khayatt and Clark went to lunch. Soon after they sat down to eat, Khayatt got a frantic call from one of his floor brokers, who tracked him down at the restaurant. The stock was getting killed. No one wanted to buy it. Short sellers, who profit when stocks fall, began making bets the offering would fail. As the lead underwriter, Khayatt's firm was responsible for supporting the stock, meaning it had to buy any shares that others sold or shorted on the first day of trading. With a new stock, investors sometimes sell if the price doesn't go up immediately, so the lead underwriter agrees to buy any unsold shares to support the price. Khayatt made a call to a friend at Loeb Rhoades & Co., a prominent Wall Street firm. Within a few minutes, Loeb Rhoades began buying large amounts of the stock. The sign of support from such a prestigious firm scared away many of the short sellers. "Loeb Rhoades really bailed us out," Khayatt recalled. Mitchell Energy raised about $10 million, but the shares closed that first day at $13.25, well below the $17 or $18 George had wanted.[7]

If he was unhappy about the sales price, he didn't have to wait long for things to change. The next year, 1973, the Organization of Arab Petroleum Exporting Countries launched an embargo against the United States and other Western nations for supporting Israel in the Yom Kippur War, sending oil prices soaring. The embargo pummeled most US stocks. Only one industry was spared—shares of oil and gas companies surged as oil prices skyrocketed from $3 a barrel to $12. By the time the embargo ended, prices would reach $30. Suddenly, Mitchell Energy and its peers were among the most sought-after stocks on American exchanges.[8] But the success was bittersweet. Oil-exporting countries in the Middle East had made it clear that they, not the United States, now controlled the oil markets. All the predictions that George and Johnny had made about the dangers of cheap imports were coming true.

George was happy for the increased value that came with the rising stock price, but once again, he was looking beyond the short-term gains. He had spent a decade studying the issue of humankind's impact on the environment and ruminating on how the planet would support an ever-growing human population. He saw other factors that few in the oil business paid attention to. Los Angeles reported its first case of smog in 1943, but it took scientists a while to determine the cause. Eventually, they found the problem: emissions from the proliferation of automobiles.[9] In 1970, Congress passed the Clean Air Act, which set new public health standards for air quality. By 1974, California governor Ronald Reagan was urging residents to "limit all but absolutely necessary auto travel" and recommending that they drive slower to reduce emissions.[10]

George saw what was happening, and while many in the oil business decried regulations that might curtail the burning of fossil fuels, he spotted opportunity. Natural gas burned cleaner than coal or oil. The way to reduce pollution, protect the environment, and preserve public health was to invest in natural gas. For George, natural gas balanced his growing interests in ecology, environmental preservation, and sustainability with his energy interests. Over the next twenty years, the company would become one of the biggest natural gas producers in the United States.

At the time of Mitchell Energy's stock offering, everyone was focused on oil prices, yet 85 percent of the company's output was natural gas or gas products from the processing plant in North Texas.[11] Few investors, it seemed, paid much attention to how Mitchell Energy's founder described the company: "We are not an oil company, we are a gas company, and gas is the fuel of the future."

◈

The truth was a little more complicated. Mitchell Energy was indeed a gas company, but it also had "Development" in its name, and that challenged investor perceptions. As he continued to study urban renewal, George also kept buying tracts of forestland north of Houston. In all, it took him three hundred transactions to piece together about twenty-five thousand acres in the configuration he wanted.[12] In the meantime, he traveled the United States and Europe studying different cities and how they developed. His real estate holdings and his growing fascination with urban renewal gradually grew into a plan: he would build a new city from the ground up—and he would do it differently than cities had been built before. He would prove that urban decay could be reduced or eliminated by designing a better community.

Thanks in part to the oil business, Houston was growing rapidly in the 1960s and 1970s, but unlike many large metropolitan areas, it had no zoning. The city had always had a strong libertarian undercurrent, believing that business could solve more problems than government. That attitude traced back to its founding at the confluence of two bayous—Buffalo and White Oak—soon after the Texas Revolution in 1836. Houston wasn't founded so much as sold. Two New York real estate promoters, John Kirby Allen and Augustus Chapman Allen, bought more than 6,600 acres that they envisioned as a hub of commerce. The Allen brothers promoted the city back East, often taking liberties in their description of the climate and terrain to entice buyers. They sold the new city as "having an abundance of excellent spring water and enjoying the sea breeze in all its freshness. . . . It is handsome and beautifully elevated, salubrious and well-watered." The first ship of new settlers arrived in early 1837, and within four months the outpost had grown to more than 1,500 people.[13] Well-watered indeed, the dirt streets frequently flooded and turned into a muddy, mosquito-infested mess lacking in salubriousness.

Nevertheless, the city grew. Over the years, Houston resisted government controls on development, and the city expanded largely at the whim of developers. George decided that his new city would evolve differently. He didn't want to build another suburb or bedroom community. He disliked the "sea of sameness" that he saw in Houston-area developments. He wanted a self-sustaining city, and his ideas coalesced around three key principles: He wanted his city to attract people from all income levels and all walks of life. He wanted it to be environmentally compatible—he hadn't bought all that beautiful forestland just to see it stripped bare. This city would reflect his belief that most people wanted to live among trees. And finally, he wanted to attract large businesses that would create jobs, making his city a place to live and work, not just a place to come home to at night.

George frequently defied conventional wisdom in looking for oil and gas, and his development proposal also stretched the boundaries of what local real estate companies found prudent. In the 1960s, everyone thought Houston's future lay to the south. The National Aeronautics and Space Administration had built Johnson Space Center—which housed Mission Control for the Apollo moon missions—in the community of Clear Lake, between Houston and Galveston. The space center attracted other white-collar businesses, such as engineering firms. The area north of Houston was riddled with creeks and rivers and was prone to flooding. Much of the land was swampy and choked with vegetation that grew rapidly in Houston's wet, humid climate.

George, however, saw something different. The area to the south was divided by small towns all the way to the coast. Growth, to the extent that it continued, would be uneven at best. While he saw white flight from the inner city as one of the biggest causes of America's urban ills, he also realized that middle-class whites tended to move to smaller suburbs—dozens of tiny political subdivisions, none of which worked together.[14] George didn't want to be at the mercy of local governments, whose visions for his city's development might change with each election cycle. The area north of Houston was either unincorporated or contained clusters of small communities and homesites. It was a clean slate on which he could test his ideas for building a better city. He began buying more land close to the interstate, building on the acreage he'd already assembled, until he amassed an area about the size of Manhattan.[15]

Just as he had seen the potential in Aspen for a popular ski development, George believed that the northern regions of Houston would become the nexus for growth in the 1970s and 1980s. The city had announced plans for a massive $3 billion international airport project twenty miles from the land he was assembling, and his property fronted Interstate 45, the main thoroughfare between Houston and Dallas, the two largest metropolitan areas in Texas.

Even before the real planning began, George had a few ideas about how he would design his city. "The first thing I'm gonna do is map out wildlife trails on the land I bought so that we don't mess up the wildlife in the city we're building," he told a friend in the YPO. "We're going to build a lot of buildings, but we're never going to build anything higher than the trees."[16]

Shaker Khayatt had no idea what George was thinking the first time he saw the land that would become George's model city. Before the public offering, Khayatt had already dealt with George's tendency to invest in pet projects. This "new city" idea seemed an even bigger obsession. Khayatt was supposed to convince investors that creating a city from scratch fit with the company's oil and gas operations. As he was driving from Houston with another company executive to see the site, the man pulled into a western-wear store and told Khayatt he needed to buy some boots. Khayatt asked why. "Because there are going to be snakes," the man said. After Khayatt put on the newly purchased boots, the two drove for miles before they finally came to an area off the freeway with a big gate held shut by a chain and padlock. "This was The Woodlands," Khayatt said. "There was nothing for miles."

The area was full of loblolly pines and live oaks, but like most of the land around Houston it was flat and positioned between the San Jacinto River farther north, and Spring Creek, which ran along the property's southern border. It was flood prone, retaining substantial water during spring rains. "It was pretty much a flat swamp," said George's oldest son, Scott, an architect who later worked for The Woodlands Development Corporation and was a home builder there.

The Woodlands may have been George's dream, but it presented a unique problem for Khayatt. Sooner or later, the project was going to need significant financing, and that meant Mitchell Energy would need to raise more money. It wasn't clear that investors who thought they were buying stock in an oil and gas company would support pouring money into real estate development, especially when, in George's mind, any return would be decades away. But it was also clear to Khayatt that George wasn't going to be dissuaded.

The original Paraskevopoulos home in Nestani. (Mitchell family)

Mike and Katina Mitchell, about 1915, Galveston. (Mitchell family)

"Uncle Jimmy" Lampis, about 1911. After George's mother died, he lived with Jimmy and his wife, Lela, for several years. (Mitchell family)

George Mitchell, about age five, with his sister Maria in 1923 or 1924. (Mitchell family)

George Mitchell, age twelve or thirteen, Galveston. (Mitchell family)

George Mitchell with his aunt, Lela Lampis, and brother Johnny Mitchell at the Lampis Bath House, Galveston Beach, about 1931. (Mitchell family)

George Mitchell, weighing about 110 pounds at age fifteen, with a 120-pound tarpon he caught off Bettison's Pier in Galveston. (Mitchell family)

George Mitchell, Texas A&M graduation, 1940. (Mitchell archives)

George (center) as captain of the Texas A&M tennis team, 1940. (Mitchell archives)

Senior year, Texas A&M. (Mitchell archives)

Johnny, George, Maria, and Christie about the time of George's college graduation in 1940. (Mitchell family)

Johnny, Christie, Maria, and George, Galveston, 1942. (Mitchell family)

Pamela Woods, George, and Cynthia Woods, Houston 1943. (Mitchell family)

Cynthia Woods, age twenty-one (wedding announcement photo). (Mitchell family)

The double wedding of George and Cynthia (at right), and Cynthia's twin sister, Pamela, and Raymond Loomis. (Mitchell family)

Figure 16. George with Merlyn Christie and Johnny in the early days of Christie, Mitchell & Mitchell. (Mitchell archives)

George at the office in the Houston Club Building, 1963. (Mitchell archives)

Fishing with his three daughters, Sheridan, Meredith, and Pamela. (Mitchell family)

Fishing with Bob Smith. (Mitchell archives)

Bernard "Budd" Clark, Mitchell Energy's vice chair, was known for his shouting matches with George. (Mitchell archives)

Catching sailfish with Cynthia in Acapulco sometime in the 1960s. (Mitchell family)

Cynthia and George at a Young Presidents' Organization meeting in New Orleans, 1969. (Mitchell archives)

Cynthia and George with their ten children at their Wickwood home, 1965. (Mitchell family)

Maria Mitchell's husband, Alando Ballantyne, with Johnny, George, and Christie. (Mitchell archives)

In the new offices at One Shell Plaza in about 1973. (Mitchell archives)

Figure 26. Mitchell Energy's first day of trading on the American Stock Exchange. Shaker Khayatt, right, with George and Charles Simpson, who worked in the Mitchell Energy public relations department. (Mitchell archives)

At the Houston Open Golf Tournament in 1976 with tennis players Andy Williams, Bobby Riggs, and Wayne Glenn. George was instrumental in laying the groundwork for Riggs's "Battle of Sexes" match against Billie Jean King a few years earlier. (Mitchell family)

Maria, Johnny, Christie, and George, undated. (Mitchell family)

Cynthia and George in front of what became the Tremont House in Galveston. (Mitchell archives)

Wise County gas plant, 1991. (Mitchell family)

At the Decatur office of Mitchell Energy and Development, 1991. (Mitchell family)

On the Strand in Galveston, 1992. (Dancie Ware)

At Mardi Gras in Galveston, undated. (Dancie Ware)

Mike Mitchell, who died in 1970 at age ninety-four. (Mitchell family)

George at age sixty-eight. (Mitchell family)

Quail hunting, 1992. (Mitchell family)

George and Cynthia at the Wentletrap Restaurant inside the T. Jeff League Building, which the Mitchells bought in 1976 and restored, the first step in a broader effort to revitalize Galveston's Strand District. (Dancie Ware)

At the Cynthia Woods Mitchell Pavilion in The Woodlands. (Dancie Ware)

Figure 39. On their Bald Head Island development in North Carolina, 1997. (Dancie Ware)

With Robert Gates at the TAMU physics building ground breaking, 2006. (Dancie Ware)

With the Lifetime Achievement Award from the Gas Technology Institute, 2010. (Dancie Ware)

Even after his knees began to trouble him, George still kept playing tennis. He's shown here with longtime tennis partners Sid Nachlas, Sherman Hink, and George Krug at the Houston Racquet Club in 1996. (Mitchell family)

With physicist Stephen Hawking. The two men became good friends late in life. (Texas A&M University College of Science)

With actor Chuck Norris, 1992. (Mitchell family)

With President George H. W. Bush at The Woodlands. (Mitchell family)

George and Cynthia with then Texas governor George W. Bush. (Dancie Ware)

George (at left) in his senior year at A&M with fellow Corps of Cadets members. Buddy Bornefeld, seated on the wall second from the left, lent George money when he ran short of funds. Later, George hired him at Mitchell Energy. (Mitchell family)

Figure 49. Katina Mitchell in about 1927, four years before her death. (Mitchell family)

Figure 50. Maria, Mike, and George on Galveston Beach in 1942 or 1943. (Mitchell family)

CHAPTER 14

How to Build a City

Tom Purcell had wondered what George was looking at as he stared at the satellite map at the end of the hallway in Mitchell Energy's offices. George wasn't just studying the land or deciding which tracts to buy next; he was envisioning an entirely new city—one that he would build from scratch. It might not solve problems such as the crime and poverty that plagued Detroit or Philadelphia, but it would set an example and provide proof that development could be done better.[1]

George gathered his ideas, as always, from a wide range of sources. He pulled from the latest research in sustainable development, from freethinkers like Buckminster Fuller, and from environmentalists such as Rachel Carson, whose book *Silent Spring* he read when it came out in the early 1960s. He was also influenced by the United Nations, which at a conference in 1972 embraced the notion of sustainable development to combat the traditional effects of economic growth and industrialization. The UN was creating a blueprint for Third World nations that would avoid the mistakes of countries that had developed ahead of them.

But George also believed that solving big problems required a business perspective. Observing the problems with urbanization during the YPO tours convinced him that the solution required a combination of social objectives and economics. "You have to make the economics work," he said. "The government can't do these projects because they don't have the entrepreneurial spirit. They don't have the management team. They have the resources, but . . . you must have that freedom of entrepreneurial spirit to make these things work."[2]

These ideas swirled together in George's mind, forming a loose congregation that gradually assembled into an unconventional development plan. George seemed less interested in making money from his real estate investments than in creating something worthwhile. "Even though he made all this money, he is not really motivated by greed," Shaker Khayatt said. He could have simply flipped his north Houston timberland and made a tidy profit.[3] With all the parcels assembled, it was now more attractive to potential buyers, and indeed, he received an offer for $80 million. He turned it down. He also could have found a partner to develop the property. He could have put in a few roads and sold off the land three hundred acres at a time, which would have generated a nice profit for the company with little risk. But he had grander ideas. By the time he'd pieced all the tracts together, George was more determined than ever that he had an opportunity to create a model community. He wasn't about to pass it up.

Even Khayatt was beginning to understand that George had a unique vision. "Once you got over the novelty of a man who thinks in decades instead of six months, you were fascinated," Khayatt said. "Here was a man who, with a straight face, told me what the world would look like 10 to 20 years in the future—the world according to Mitchell." He would lay out a vision of what Houston would be at the end of the century. "Listening, you start by being somewhat suspicious—is this a fairy tale?—and then very quickly you come to the conclusion that it is real, and you really have met somebody who is extraordinary. He also had a sense of morality that these days is hard to find."

George's new town would be a master-planned community, incorporating the latest theories of design, sustainability, and livability. What makes a city livable? George scoured development, environmental, social, and real estate studies. He found an interesting contradiction of human nature: most people say they want a rural lifestyle, yet they also want the amenities and conveniences of a big city.[4] George set out to prove they could have both.

He wasn't alone. In the years after World War II, developer William Levitt built a new community in Pennsylvania for returning veterans and their families. Levittown was the first true planned community, but its houses were built using assembly-line techniques that resulted in a uniformity that many would later find stifling. Now, a new generation of developers was expanding on the idea of what a planned community could be. At a conference of like-minded planners, George met Dick Browne, who had spent a decade as a land surveyor and later as the mayor of Wayne, New Jersey, before joining Rouse Co. in developing Columbia, Maryland, between Washington and Baltimore. George told Browne he'd assembled a parcel of land north of Houston and asked what it would take to do a development like Columbia. Browne replied: "You'll need lots of land with good highway access near a big city; extensive land surveys; engineers for planning drainage, water and sewer systems, transportation and road-building; architects and landscape architects; finance and accounting expertise; marketing and sales people; patient and plentiful capital resources; a dedicated skillful staff; environmental sensitivity; an upbeat expanding market; and knowledge about building a human settlement."

"Is that all?" George shot back.[5]

Rather than remember the whole list, he hired Browne, and they studied what made different areas appealing. "We looked for spots that were uplifting to the human spirit—places in Europe like the Tivoli gardens of Denmark, sidewalk cafes in Paris, the old piazzas in Rome," Browne said. "In the United States, we saw Ghirardelli Square in San Francisco, Georgetown in Washington, DC, and the Riverwalk in San Antonio, Texas."[6]

The idea of building a new town wasn't new. Ebenezer Howard, a British urban planner, pioneered the garden city movement outside London in the late 1800s, a response to the dense squalor that came with the Industrial Revolution. Until then, cities had grown from settlements, usually situated near a river or other source of transportation. They grew haphazardly as more people moved there. Howard pioneered the idea of planning the development. He proposed water and sewer lines from the central city to his new suburbs, and rail lines that would move people into the city for work. George traveled to England to study Howard's designs.

From a business standpoint, it was a big risk. If the project failed, it would be a total loss for the company. George decided to go ahead with the concept for one simple reason: he wanted to give something back to Houston because of his own good fortune.[7] Like many immigrant children, he never forgot that his illiterate father had left the Greek mountains as a goatherd and come to a country where his children could become educated and wealthy.

For all his big dreams, though, the biggest hurdle to the new city project was a familiar one for George—finding the money to pay for it. Neither he nor the company had the tens of millions of dollars needed to fund it.[8] George built his energy business on debt because it was the only way he could capitalize on the Boonsville Bend and other discoveries. But now, the company's heavy debt load worked against him. Few banks would lend an energy company millions to develop a city in the middle of nowhere, and while some were willing to consider the land itself as collateral for development loans, they demanded onerous provisions. The banks either wanted George to guarantee the debt personally or they wanted the right to pull their financing at any time. That meant if the project failed, George and the company could be wiped out. If the banks decided to call the notes, they could use the liens against the property to take over the entire project.

George told Jim McAlister, his financial fixer, to come up with alternatives. McAlister considered setting up a joint venture that would give a partner partial ownership of the development. But ever since George had bought out Merlyn Christie, Bob Smith, and his brother Johnny, he had avoided partnerships. He didn't like sharing control of his projects. Even the stock structure of Mitchell Energy reflected that. The dual class structure of the stock, with its A and B shares, prevented his investors from dictating returns. Without that structure, his new city wouldn't have been possible.

While McAlister worked on the funding, George pushed ahead with his plans. He hired Houston architect Cerf Ross to begin designing the new community, and in a stroke of luck, Ross had an idea for solving the financing problem. He had been working on Native American housing projects in Arizona and New Mexico, and he told George about a new federal program that might help.[9]

Congress created the Department of Housing and Urban Development as a cabinet-level agency in 1965 to address the lack of affordable housing that had contributed to the riots in some of the cities George had toured with the YPO. In 1970, lawmakers passed legislation that established a slew of new loan programs for housing initiatives, collectively known as the New Community Program, or simply Title VII.[10] One of the measures provided $50 million in loan guarantees for developing a new town. George was reluctant to enter into a financial arrangement with the federal government, but he didn't have better options. The HUD program came with big strings attached—requirements for affirmative action in hiring, affordable housing for low-income residents, and lots of bureaucratic red tape. But the government initiative would allow George to retain greater control over his project than bank financing and would remove many of the financial uncertainties.

Besides, George was no stranger to tapping pools of federal money. In the coming years, in both his real estate and energy holdings, he would use an array of tax credits for agricultural exemptions, mobility, and conservation. "He was a master at leveraging entities—any public entity—to benefit his business or whatever project he wanted to carry forward," his longtime publicist Dancie Ware said. Nothing would put those skills to greater use than building his city.

In February 1970, McAlister and Ross completed a proposal for HUD, using the name "Satellite City" as a placeholder. The two booked a 7:00 a.m. flight to Washington to present their plan. McAlister's phone rang at 2:00 a.m.—five hours before the flight. It was Ross. He'd been going over the final touches on their plan and he'd spotted a statement, buried in HUD's loan guidelines, that the agency didn't want to finance satellite cities. It sought self-sustaining developments rather than future bedroom communities that relied on their proximity to larger metropolitan areas. HUD didn't want to underwrite the same sort of white flight that had contributed to the country's urban woes. "We can't use 'Satellite City' as a placeholder name," Ross said.

They kicked some names around. It was a forest. There were lots of trees. Woods. They slapped the name "Woodland Village" on the proposal.[11] The document was due in a few hours, and it was the middle of the night, so they didn't bother to run the name by George. After all, it would probably be changed later anyway.[12] In recounting the history of The Woodlands, George told a different version of the naming. Among the timberland in his original purchase in the early 1960s, he had set aside about five thousand acres near the town of Magnolia as a private family retreat, which Cynthia dubbed "The Woodlands." She liked the name so much she suggested they use it for the new city. (The retreat, later renamed Cook's Branch after the spring-fed creek that runs through the property, is still in the Mitchell family.) It's not clear whether George or Cynthia ever saw "Woodland Village" on the proposal, or whether the similarity was a coincidence. Regardless, "The Woodlands" became the new town's official name.

By June, McAlister had heard back from HUD officials. They were intrigued by the proposal, and they didn't want to discourage submissions to their loan program, but in private they told him the plan wasn't sophisticated enough. If George and his team wanted to be taken seriously, they needed to step up their urban planning skills. George, of course, didn't have any urban planning skills, and neither did McAlister.[13] McAlister was a financial fixer and George was a dreamer who, like Buckminster Fuller, often glossed over details. Mitchell Energy was still a company that followed the wandering path of George's interests, and employees often found themselves tackling tasks for which they had no particular aptitude simply because George had confidence in them. Unlike Fuller, though, George had the money to acquire the expertise he lacked. He hired Houston planner Robert Hartsfield, who was struck by George's sense of the natural world and how it should interact with design and development. Hartsfield recommended that George read *Design with Nature*, a book by Hartsfield's college professor, Ian McHarg. McHarg chaired the University of Pennsylvania's landscape and regional planning department and was a pioneer in the environmental planning movement. He believed that landscape architecture was as import as building design and engineering and promoted harmony between humanity and nature, rejecting the "self-mutilation" of modern development. Humanity, McHarg wrote, "must become the steward of the biosphere," which would require us to "design with nature."[14]

George was so impressed with the book that he hired McHarg

as a consultant to help plan The Woodlands. During the next eighteen months, McHarg developed an environmental master plan that preserved 25 percent of the native vegetation, and he compiled an inventory of the trees.[15] Thirty percent of the land was set aside for greenbelts, parks, golf courses, and lakes. The main roads were designed as parkways with seventy-five-foot greenbelts so that people driving through town would see only trees, not buildings. Rather than have them lined with commercial developments, as in so many areas of Houston, the commercial centers were concentrated in nodes at key intersections. Fewer businesses on the main roads meant fewer curb cuts, which in turn meant the thoroughfares could handle as much as 30 percent more traffic than roads lined with businesses and shopping centers.[16] That increased traffic capacity was important for commuters who needed quick access to Interstate 45 and downtown Houston.[17] In keeping with George's method of thinking, everything was planned with an eye toward the distant future. The design projected 160,000 residents, and as of 2018, The Woodlands still hadn't reached that size.

"There was not much left to chance. Everything about that was incredibly well thought out," said Jim Blackburn, the environmental lawyer who worked on the project. Businesses were not allowed to have neon or internally lit roadside signs, which at first made it difficult to attract restaurants because most of the major chains had specific sign requirements. George personally negotiated with McDonald's Corp., getting it to agree to the sign restrictions even though the company's franchise requirements called for a large neon sign with the traditional Golden Arches. McDonald's also wanted to be within sight of the freeway, which violated George's vision. He convinced the company to tuck its store into a village center off Woodlands Parkway, the main thoroughfare. That way, he pointed out, the restaurant would be surrounded by residences. George wanted commercial centers where locals would shop, visit the doctor, and eat. He didn't want people running out to the freeway for everything. Within a few years of opening, The Woodlands' McDonald's was one of the highest-grossing locations in the Houston area.[18]

Although the main thoroughfares started as two lanes, the plan called for them to be expanded as the community grew. Because George owned the land, the city would have no right-of-away acquisition costs in the future, keeping development expenses low. This, too, was a change from the status quo. Most development around Houston at the time was built off two-lane county roads. As the development grew and the inevitable strip shopping centers sprang up nearby, traffic became a problem—but not one the developers had to worry about. That was left for the residents and county officials to sort out. The result was chaotic suburban growth. By planning the roads as well as the commercial development, George believed The Woodlands would avoid the randomness and choked infrastructure that plagued so many other parts of greater Houston.[19]

Built back from the main roads, the side streets twisted and turned, sometimes looping around on each other or ending abruptly in cul-de-sacs. New residents would find the patterns confusing, but the strategy was designed to reduce crime because thieves, too, would be thrown off. It was harder to make a getaway in a strange area if the roads weren't laid out in a typical grid. The unusual traffic patterns also kept residential areas free from through traffic. Because the shopping mall, other retail areas, and office buildings were located either close to the freeway or in village centers, nonresidents didn't have to drive through neighborhoods to get to work or stores.[20] McHarg also conducted a hydrological study that became the basis for flood protection. This was vital because the Houston area is only a few feet above sea level and prone to heavy spring rains.

McHarg also conducted an unprecedented environmental survey that studied both the soil and vegetation and then tried to match land use to the areas best suited for it. Regions with heavier clay soils that absorbed less rainwater were pegged for the densest commercial development. Areas in which soils were more porous would feature the lighter footprint of residential neighborhoods.[21]

The revised proposal worked. McHarg's analysis impressed HUD so much that the agency used it as a model for other environmental impact studies in government-funded developments. George was satisfied, too. He saw too many developments around Houston with arboreal names like "Oaks" and "Pines" in which the developers chopped down all the trees. Clear-cutting each lot made drainage easier because builders could slope the land however they wanted. George was adamant that The Woodlands would, in fact, have woods. "You have to fight me to cut down a tree or cut the wildflowers," George would say later. "And that's what makes The Woodlands interesting. It's not just a bunch of lawns."[22] Lots—which maintained many of the trees and much of the native vegetation—were designed with a complicated drainage system that preserved the natural forest environment and at the same time reduced the risk of flooding.[23]

Essentially, George turned the development process on its head. Most developers look at raw land and figure out how to remove the forest to make room for the people. Instead, he focused on inserting 160,000 people among the trees while disturbing the land as little as possible. The Woodlands slogan would become "the livable forest," and to George it was more than just marketing hype. He brought in biologists from his alma mater, Texas A&M, who identified two hundred plant species native to the area. The scientists recommended leaving 600-foot-wide corridors of green space throughout so that deer and other wildlife could reach water sources such as Spring Creek and Lake Woodlands, the water feature at the center of the development. George's insistence on saving every possible tree meant that for many years The Woodlands was a haven for the red-cockaded woodpecker, a shy and endangered species that lived in the area.[24] (The woodpeckers' habitat has shrunk over the years, but the birds still nest in a state-protected forest on the northern edge of The Woodlands, as well as on the Mitchell family's Cook's Branch preserve.) Drainage ditches ran through the development, and unlike the bayous in Houston, they were not paved with cement.[25] Although the land was mostly flat, the few slight ridges were selected for the major roadways. This, too, would help with drainage, because water would flow away from the roads and toward the network of ditches.[26] Porous concrete was used for parking lots to reduce rainwater runoff.

McHarg included additional ideas, some of which came from George and others. The Woodlands would have no gated communities, and it would be divided into six villages that functioned as separate neighborhoods, each with its own shopping and commercial center. Each village would be home to about eighteen thousand people, and the houses would be clustered, mostly on cul-de-sacs, that would keep neighborhoods to about thirty families each. Pedestrian trails linked the villages, each of which was designed to attract residents of all income levels and ethnic backgrounds.[27] The trails were laid out to ensure that children could walk to school without crossing a major street.[28] People could walk to the shopping centers and offices, which were sheathed in mirrored glass. When asked why, George replied: "So the deer don't see the office buildings."[29]

George typically didn't spend much time on details, but when it came to planning how his city would evolve, few were too small for his attention. Schools and churches would share parking lots, because they used them on different days of the week. Churches of different faiths could share facilities, too, further reducing the environmental impact—and expense to the congregations—of maintaining separate buildings. He called the concept Interfaith. At a coffee hour in the parish hall of St. Francis Episcopal Church near George's home in Houston, he began discussing his plans for The Woodlands with the Reverend Richard Wheatcroft. George told him he wanted a creative idea for addressing the land use and operating costs of churches.[30] George had served on the parish board and had seen how much the church spent on utilities and facilities management for a space that it used essentially once a week. He wondered why three churches of different congregations didn't share one campus like a university.[31] He hired Wheatcroft to spearhead the effort. It turned out churches didn't share space because each wanted its own. Rather than a shared space, Interfaith would become an incubator for new congregations. (Today, it serves as a sort of joint charity and community services organization, operating childcare facilities, a food bank, and services for seniors.)

The idea for the villages came from George's childhood memories of Galveston. As a city of immigrants, Galveston evolved around ethnic neighborhoods, each with its own street-corner shops. Young George would wander from one neighborhood to the next, experiencing different foods and cultures, and he wanted to re-create that feel of connected communities within his new town.[32] "He saw first-hand the human potential of people of different origins living together," George's son Scott said.

The design of The Woodlands was George's rebuttal to everything he disliked about development in Houston, from the lack of trees and parks in the Memorial area that so frustrated the family when they first moved there to the plans to cement Buffalo Bayou that he fought in the 1960s.[33]

As the plans progressed, George insisted that the company cut the price of certain lots so that affordable housing was included in the mix of home building. When The Woodlands opened in 1974, it included plenty of pricey homes on golf course lots, but it also included new homes that sold for as little as $28,000, the equivalent of $146,000 today.[34] To prevent the houses from all looking the same, as they did in many tract developments around Houston, George brought in twenty builders and encouraged them to try different designs—not just for expensive custom houses but for starter homes as well.

Attracting upper-middle-class buyers was no problem, despite the distance from downtown Houston. George knew that golf courses would draw wealthier homeowners, but even here, he broke with tradition. Many of the original homes on The Woodlands' golf courses were decidedly affordable.

While he recognized the importance of families in building his city, George knew that a community of only single-family homes would turn The Woodlands into another suburb. He didn't want to replicate planned developments like Columbia, Maryland, or Reston, Virginia, which he saw at the time as bedroom communities. He wanted a city where people lived and worked.[35] He envisioned a waterway that would flow through the town center, served by water taxis that moved people among the shops and restaurants. Part of the waterway would be lined with multifamily housing to attract young professionals with modern accommodations close to amenities such as restaurants, theaters, and concert venues.[36]

Cynthia, too, wanted the community to reflect an appreciation for the arts that were available to everyone. After attending a performance of the Austin Ballet with her daughter Sheridan Lorenz at the Zilker Hillside Theater, Cynthia marveled at how many people of all ages, from babies to senior citizens, were sitting on blankets watching the performance. It had always bothered her that access to events like the symphony was limited largely to the social elite or those who could afford the ticket prices. The Austin performance, she told Sheridan, reminded her of Russia, where everyone had a chance to enjoy the performing arts. "We should do something like this in The Woodlands," she declared. The idea eventually became the Cynthia Woods Mitchell Pavilion, now the most successful outdoor performing arts venue in the country. (The pavilion is the "summer home" of the Houston Symphony, which puts on free concerts.)

The bigger challenge for The Woodlands, though, was luring lower-end home buyers to a community that had an upscale feel and was more than thirty miles from downtown Houston. "He really wanted an ethnic mix in The Woodlands, and it's the one thing that really didn't quite work for him," Scott Mitchell said. "We didn't have the blue-collar jobs nearby that Dad really wanted."

Despite these good intentions, it remained to be seen whether a collection of well-off white men could create the economic and racial diversity that George desired for his town. "It was something both my parents were passionate about," Scott Mitchell said. "They worked with all kinds of people. I think Dad was universally popular with everyone who ever worked for him. He was a man who didn't see color." Yet his own company, in both energy and real estate, was run almost exclusively by white males.[37]

Just as George didn't incorporate his ideas of sustainability into Mitchell Energy's practices, he also didn't infuse his company with his progressive ideals of race. The oil business, even in the 1990s, had a dearth of diversity. Yet for George, it had less to do with industry practice than with the way he compartmentalized his interests. Mitchell Energy may have owned The Woodlands, but that was just a financial arrangement. The two divisions were separate entities. George's interests in sustainability, historic preservation, and later, physics and astronomy were all pursued separately as well.

With McHarg's help, George managed to impress HUD officials

with his vision for how the Houston area would grow northward, but there were few guidelines for what came next. A few cities—such as Columbia and Reston—had been built as planned communities, and HUD had recently approved a proposal for Flower Mound, near Dallas, but George had few examples to draw from, especially for his vision of incorporating environmental sustainability into his development. "There's no roadmap for what we're doing," McAlister said. "There aren't many experts available to say, 'come and do this for us.' We're out there by ourselves."

CHAPTER 15

A House Divided

In its early days, The Woodlands suffered from the same lack of spending controls as George's other projects. Mitchell Energy kicked in $5.5 million, and George borrowed $50 million against the HUD loan guarantee. The money went quickly. The first $2 million covered the financing costs. Another $23 million paid for the mortgages and liens on the land he'd assembled for the project. George didn't have a choice on this. HUD would guarantee the loans only if the development had clear title to the land should the project fail. Setting up The Woodlands Development Corporation, which would oversee the enterprise as a subsidiary of Mitchell Energy, cost another $3.6 million. That left about $27 million for the new town's infrastructure—roads, utilities, parks, trails, and other amenities.[1]

The project, though, had a significant shortcoming. George and his staff lacked the expertise to implement his plan. He had developed a couple of residential communities in Galveston and Cape Royale on Lake Livingston, but he had never done anything on the scale he now planned. So far, most of the people working on it had been reassigned from the oil and gas business and pressed into service on The Woodlands. George didn't care much for credentials. He tended to judge employees on their merits, by what they could do, but if The Woodlands were to succeed, it would require people who knew how to run a large-scale development company.[2]

Rouse Co. was a private developer that built Columbia as a 14,000-acre "color-blind," aesthetically pleasing community. While the projects differed in some ways, Dick Browne and others at Rouse had been helpful in offering advice when George put the

HUD proposal together, and he knew many of the executives there. He began recruiting them, as he had with Browne.

George, McAlister, and other team members went to Maryland and met with company founder Jim Rouse and others to understand how they had built their cities. One disturbing fact emerged: all the developments had financial problems. In Columbia's case, the villages were built too far apart, which increased infrastructure costs, and the company wasn't selling lots fast enough to keep pace with expenses. George decided that for The Woodlands to be financially viable, it needed to grow in concentric circles, radiating outward to keep costs in check. This would also inspire a sense of community that would start small and build with the growth of the development.[3]

As George hired more people from Columbia and moved them to Texas, the group began to function as an autonomous unit. The team spent money with little regard for overall costs. Some Mitchell Energy employees saw them as arrogant and began referring to as them as the "Columbia Mafia." "The real estate people came in, they were a different breed," said Joyce Gay, who had been George's first assistant and by then was a vice president. "We were a house divided after that."

On the energy side of the business, many of the employees had worked with George for years. They had helped the company grow and enjoyed strong personal relationships with each other, the firm, and George himself. Many resented the amount of money The Woodlands consumed.[4] "We felt like we were making the money and they were spending it," said Howard Kiatta, the geologist. Budd Clark and other top executives also expressed displeasure at the expenses.[5]

The real estate side of the business was populated largely with outsiders who brought with them the office politics and one-upmanship that are typical in many work places, but that clashed with the familial atmosphere of Mitchell Energy. Above it all was George, who on both sides of the company was larger than life. Everyone wanted to curry favor with him.[6]

While the Columbia recruits had the experience George needed to make his project a reality, many didn't fully understand his commitment to the environmental aspects of The Woodlands.[7] "They came in with the attitude that they were there because we couldn't do it," McAlister said.[8]

George could fixate on the tiniest detail—saving a particular stand of trees or obsessing over the design of a fountain—but he wasn't good at delegating managerial tasks.[9] His mind could see the relationship between his overall vision and those personal touches, but he didn't concern himself with how building the details into the project got done. He was out of his element in running The Woodlands, and perhaps the Columbia Mafia sensed that. Or perhaps they simply weren't used to a man who was so focused on his vision of the future. Either way, they took liberties with the latitude he gave them.

They threw lavish parties. Jim Blackburn remembers more than once finding their offices littered with wine bottles after lunch. "George trusted a lot of the guys who came in from Columbia, and they took advantage of him," Blackburn said. Just as he had done with his father and his brother Johnny, George could overlook the flaws of people who worked for him. When other managers pointed out that an employee wasn't performing well or wasn't trustworthy, George would say, "Well, you just have to watch him," or "We just need to kick him a bit." He once fired a hotel manager in Galveston for stealing, only to suggest rehiring him a few months later because the man had been a better manager than his replacement, except, of course, for the theft.[10]

When Clark complained about the expense account of Len Ivins, the leader of the Columbia group and the first president of The Woodlands, George replied: "Budd, that's the cost of having him here. I'm willing to pay for it, so shut up." Eventually, though, the board lost patience with Ivins's freewheeling ways and his running tab of champagne and steak dinners and ordered George to fire him.[11] "Dad always lamented losing him," Scott Mitchell said. "He would say 'Len was creative.'"

Time and again, George showed unwavering loyalty to his employees, just as he had written faithfully to Bryan Smothers, the photographer, in Vietnam. Out of one hundred employees, perhaps two worked against him, but to George that 2 percent could be overlooked because the loyalty from the others made up for it. "Dad was loyal to 100 percent of his employees," Scott said. "And 98 percent of them gave him 110 percent back. That's why the formula of trusting everybody worked for him."

Cynthia was less understanding. She clashed with the Columbia Mafia and other Woodlands executives over public art projects and the choice of artists and architects. When she became frustrated, she would tell George: "Get rid of the bums! Clean house." At Cynthia's urging, George had agreed to set aside one-quarter of 1 percent from each commercial land sale for public artwork, such as sculptures placed in front of the village centers, an initiative that became known as the "art pot." (By 2005, more than $3 million would be collected to fund and maintain public art projects.)[12] Cynthia wanted the architect Michael Graves to design a small theater for about six hundred people, but The Woodlands executives felt the project was too costly and too limited in its public benefit. They went to George and convinced him to scuttle the plan, which infuriated her.[13]

◆

With the initial funds depleted and the Columbia Mafia spending money on everything from fancy dinners to amenities for attracting new home buyers, The Woodlands quickly ran into cash flow problems. It enacted a policy of paying small contractors first, while trying to hold off larger ones. Worse, by the time the development opened in 1974, no one in the company really knew how much had been spent.[14]

The staff for the project had ballooned to 365 people plus a host of outside consultants and engineers—at one time, as many as fifteen different engineering firms were working on design elements—and they often competed at cross-purposes and wound up wasting money.

George simply didn't have the cash to do all the things he wanted to do, but he wouldn't let that deter him. Nor did he seem bothered by other obstacles that threatened to keep The Woodlands from succeeding. Even the weather, in both summer and winter, seemed to conspire against the development, delaying construction in the years leading up to the grand opening. The area north of Houston typically receives about forty-nine inches of rain a year, but in 1972, it got ninety-six inches. Known for mild winters, the area also experienced two rare snows that year.[15] HUD caused further delays when it balked at deed covenants that would require residents to pay fees to community associations for recreational amenities.[16] The Woodlands wasn't an incorporated city and it had no taxing authority, so homeowners' fees were the only way to get citizens to pay for services.

The status pushed back the opening of The Woodlands from January 1974 to August and then to October, when twenty-two thousand people turned out to see George's vision firsthand. He gave a big speech, of course, as did US representative Charlie Wilson, who a few years later would become famous for aiding Afghan fighters in their efforts to repel the Soviet invasion of Afghanistan. Other festivities were farther afield. Buster Crabbe, of *Flash Gordon* fame, and *Star Trek*'s William Shatner made appearances. There were archery exhibitions and live music that included country and western, mariachi, and Dixieland bands.

On January 1, 1975, the first family moved into a townhouse in George's new city. Within a few months, dozens of houses were occupied. The opening fanfare, though, masked The Woodlands' mounting financial troubles. After his election in 1972, President Richard Nixon canceled the government's new towns program, suspending all additional funding. George had been counting on another $33 million from HUD, and he considered suing the feds for breach of contract but decided against it. Part of the reason Nixon opposed the program was that all thirteen cities for which HUD had guaranteed loans were near bankruptcy.

Soaring gasoline prices following the 1973 Arab oil embargo raised another worry: many potential home buyers might reconsider living twenty-seven miles from downtown Houston. George tweaked his plan to add more commercial development—increasing the likelihood that people could live and work in the community—and he set up a commuter bus service to central Houston.[17] But prices weren't the only threat to The Woodlands' future. In the wake of the oil embargo, the country slid into a recession, which curtailed mortgage lending and ate into home sales. Interest rates soared to their highest in forty years.[18]

Once again, The Woodlands turned to Mitchell Energy for a

bailout. The company advanced the real estate side of the business $15 million, but even that wasn't enough to cover the shortfall. George had largely deferred to the expertise of the Columbia Mafia, but by early 1975, he could see that something had to be done to bring greater economic stability to his project. He started by slashing the staff of The Woodlands Corporation to 160 from 365—including many of those who'd been hired from Columbia.

Ironically, the same energy crisis that was hurting home sales in The Woodlands helped save it. Soaring oil prices meant that Mitchell Energy's reserves were now worth more than five times what they had been. The Woodlands may not have had any cash, but George now had collateral to borrow millions more to keep the project afloat, even though he had to pledge his ownership in the energy company to HUD to seal the deal. George's dream was teetering, but it was still alive. As one top real estate executive put it: "The Arabs saved The Woodlands. If we had not had that oil embargo, we might not have survived."[19]

The increased value of the oil assets enabled George not only to get The Woodlands through its cash crunch, but to expand by buying more acreage around the development as it became available. He was so determined to make the project work that he warned his family that he would dedicate his personal wealth to it if he had to.

As part of his vision to make The Woodlands more than a bedroom community, George wanted to attract jobs to his new city, and he intended to lead by example. In 1980, Mitchell Energy moved out of One Shell Plaza in downtown Houston and into a low-rise office building—its top floor well below the tree line—in The Woodlands.[20] George set up a mortgage company to offer low-interest loans to employees, and hundreds of them moved to the new city. Many, now retired, never left. That not only strengthened the relationship between the developer and the citizens but also created a shared sense of commitment. "We were not absentee owners," said Roger Galatas, a geologist who had worked in Exxon's land development business before joining the Mitchell team and eventually becoming president and chief executive of The Woodlands.[21] George moved there too, leaving his beloved Wickwood home for a far more modest, 4,100-square-foot wood frame home designed by his son Scott.[22]

When the first residents began moving to The Woodlands in October 1974, they entered a town that was fundamentally different from a typical American suburb.[23] It wasn't just that the community was planned; it was how the plan—the roads, shopping centers, and environment—all came together. The Woodlands, in many ways, was an extension of George's personality. A more egotistical developer might have run the town as his own fiefdom or as a monument to himself, but George's presence was felt more in the details. He fretted over the aesthetics.

When he realized that residents didn't have medical care nearby, the company paid for the first hospital. It even considered setting up its own phone company.[24] As construction began and the first areas were cleared for houses, George would often stand and watch, making sure the crews didn't cut down the biggest trees. He demanded that the fountain in front of an office building be remade three times until it looked the way he thought it should.

The new residents weren't even aware of some of the touches that George and his team had incorporated into The Woodlands' design. They saw the efforts to preserve the forests, of course, but they didn't necessarily appreciate that it was more than a beautiful setting. The greenery was also a natural barrier against floods from heavy rains, tropical storms, and hurricanes blowing in from the Gulf of Mexico.[25]

There was another characteristic many residents didn't see. Mitchell Energy owned all the commercial buildings. Just as George believed in holding on to oil and gas leases, he believed developers should retain the buildings they built. It was unusual, even in master-planned communities, but George believed it kept the company more involved in the community and more aware of residents' needs. The firm wouldn't make much money on leases, but in the long run, the strategy would pay off in a more attractive and successful community, he argued. "He was patient; he didn't worry about short-term results," Galatas said.

George's innovation had become more than a personal success. It served as a model for suburban development worldwide. Even in its early years, The Woodlands drew city officials from across Europe, the Middle East, and Asia, as well as the United States, who wanted to study George's design.[26] Master-planned communities became the suburbs of choice, though few matched the amenities and attention to detail that George instilled in The Woodlands.

Most developers couldn't justify the cost—or the financial losses—that George endured in bringing his dream to fruition. Over the years, he found financial Band-Aids to keep the project going, but The Woodlands never made money for him. In fact, the project cost the company $20 million to $30 million a year. As was so often the case, though, George didn't seem to care. For him, The Woodlands was about creating something new, about implementing a better way of building a community. To him, that pursuit mattered more than money.

The Woodlands lurched along, caught between economics and environmentalism—all governed by HUD's looming bureaucracy. George wasn't worried about a short-term cash crunch, and he remained confident that his city would be a moneymaker over the long term. Besides, he had paid an average of $1,688 an acre for the twenty-three thousand acres that made up the original development, and by 1981, he was selling lots for $62,000 an acre.[27] But the project lived in the shadow of the financial failures of other "new cities," and HUD demanded more and more economic safeguards to protect its loan guarantees and instituted overbearing requirements for everything from environmental protection to ethnic diversity. George's earliest fears about losing control by partnering with the government were coming true. In 1981, he decided to pull out of the HUD program. The move gave The Woodlands some economic wiggle room, which may have been the reason the community survived and later thrived. Of the thirteen Title VII programs HUD funded, only The Woodlands repaid its loans without requiring a subsidy from the federal government.[28]

CHAPTER 16

Galveston Reborn

Coulson Tough took pride in what he'd helped put together with The Woodlands. There was still much to be done—George used to talk about what things would be like in fifty years—but Tough knew they had built the foundation for a new type of city. Tough had joined the staff of The Woodlands Development Corp. in 1973 after working in land acquisitions for the University of Houston. George quickly put him in charge of developing roads and utilities. Now, with residents moving in, those issues were largely resolved, and with the staff reductions that had followed the opening, Tough waited to see what George would have him do next.

The call came, as they so often did, on Sunday night. Like many Mitchell Energy executives, Tough kept paper and a pen near the phone because when George called, he often issued rapid-fire orders. He also had a habit of giving the same orders to more than one person, and nobody wanted to be the one who didn't understand what he wanted. "We're going to do some historic renovations in Galveston," he told Tough. "There's some tax benefits. Go to Galveston and buy a building." Then, he hung up before Tough could ask for specifics.

George had moved to Houston in the mid-1940s, and to The Woodlands in the 1980s, but as much as he loved his new city, his heart remained on the island of his birth. He and Cynthia still had the weekend home on the beach, built in Pirates Cove in 1978—and he frequently met childhood friends for coffee in the Kroger grocery store on the Seawall. "I was born in Galveston, and I love it better than any city in the world," he told the *Galveston Daily News* in 1969.[1]

In his mind, few other places measured up. For twenty years, Cynthia worked on convincing him to take a cruise to Alaska, and when George finally agreed, he was miserable. The ship had no tennis courts and no telephone. He was completely cut off. "He was like a caged lion," his oldest son, Scott, recalled. George didn't do leisure time. He didn't understand that the point of a cruise was to relax and enjoy the scenery. "All we did was eat and sit around," he complained. But Cynthia had wanted to see the glaciers, and as the ship pulled within sight of the Alaskan coast, she took him up on deck. Standing side by side, they looked at the breathtaking view of an ice floe shearing off the side of a cliff.

"My God," Cynthia said, "isn't that beautiful?"

"Well, it's beautiful," George replied, "but it's not Galveston."[2]

The Lafitte's Gold discovery had helped Galveston economically, but George wanted to do more. His hometown was struggling. The Strand had deteriorated from the "Wall Street of the Southwest" into a rundown stretch of abandoned buildings and warehouses. The Seawall was mostly a string of cheap motels. Its days of celebrities and gambling were decades in the past. By the early 1980s, Galveston was a hangout for low-rent beach bums. The local joke was that tourists came for the weekend with a ten-dollar bill and a dirty shirt and didn't change either one.[3]

For George, the decay of the city coincided with the loss of his own family connections on the island. His father, Mike, died on Christmas Day 1970 at age ninety, after years of declining health.[4] He'd had a leg amputated, he was almost blind, and he'd been dogged by persistent heart trouble. He'd been in the Moody House nursing home for several years, but he'd still managed to chase women and make deals, and he'd continued to get letters from friends who wrote to him from around the world. For years, Mike had befriended not just island regulars, but also dozens of seamen passing through the port, many of whom would return years later to visit him.[5]

Mike was cremated wearing a red tie and a black silk suit with a carnation in the lapel. The family buried some of his ashes with Katina and sprinkled the rest over the site of the Snug Harbor Cafe, the establishment a block from the Seawall that had been a pressing shop, restaurant, family home, and Mike's base of operations for decades.[6] (After the fire that destroyed Snug Harbor, Mike lived in different apartments before moving to the nursing home.) George's oldest brother, Christie, died a decade later, on February 28, 1980, at age sixty-seven.[7] His ashes, too, were split—some were buried with Katina and Mike and the rest were scattered on the beach he loved. George continued to own and operate Christie's bar, the Beachcomber, for years after his brother's death.[8]

But he wanted to do more. George's directive to Tough was the beginning of his effort to transform his hometown. His plan went beyond his brother's bar or his Pirates Beach development, or even Lafitte's Gold. For one thing, George was putting his personal money—not the company's—into the project, and for another, he planned to single-handedly lead an effort that would restore Galveston's past glory, minus the bootlegging and gambling. Cynthia, too, was deeply involved in the project. From the time she had moved to the island after she married George, she was appalled by the city's demolition of beautiful old buildings that had fallen into disrepair.[9] Galveston had more than two thousand Victorian structures, one of the highest concentrations in the country, and George and Cynthia decided they wanted to save as many as they could.[10]

Tough did as he was told. When George called to ask about the trip, he said he'd bought the Thomas Jefferson League building, a dilapidated, three-story, Renaissance Revival–style commercial structure built in 1871. The brick building featured a cast-iron facade from New Orleans and a huge cornice that the *Galveston Daily News* declared "an ornament to our city" when it opened.[11] The building changed hands over the years, home to clothing shops, a savings and loan, and offices for insurance agents, attorneys, and a steamship company. It ended up as a wholesale hardware outlet, which closed in 1972. A year later, the Galveston Historical Foundation purchased it through its Strand Revolving Fund, which it set up to buy historic buildings and resell them to developers who would restore them.[12] Despite growing up in town, George didn't know which building it was. Tough told him it was at the corner of Strand and Twenty-Third Streets. "Oh, OK," George said. "Let's get going." And he hung up.[13]

For George and Cynthia, the restoration project was a chance to work together in a way they hadn't even in The Woodlands. Cynthia had never been content to play the doting housewife of an oil millionaire. The two shared a deep mutual respect, but as they aged, the marriage was fraught with conflict. Outwardly, they appeared as devoted as ever. George still insisted on including Cynthia's name on their philanthropic projects, and he would frequently tell his children that their mother "was a wonderful woman." Cynthia, for her part, never seemed to care that George made a lot of money.[14] Instead, she seemed to relish the ideas they shared.

But those ideas often clashed when it came to The Woodlands. Cynthia felt that George was too quick to listen to his executives, who often overruled her. The mutual respect they projected outwardly masked a deepening discord. The oldest children recall their mother and father showing affection toward each other, but their younger siblings describe a growing distance between their parents. While each maintained a good relationship with their children, they seemed to have a harder time maintaining their commitment to each other. Having each grown up in single-parent households, neither had role models for how to be a supportive spouse. The outward appearance of marital harmony masked the absence of an emotional connection. "My father didn't have a great ability to communicate personal or intimate thoughts, and we all understood that," Todd Mitchell said. "It wasn't a problem if you understood who he was and what he felt. You kind of had to decipher some of that, and you just accepted it."

George could be private, distant, and even hypocritical. Cynthia's interest in preservation and the arts put her in touch with many architects and artists over the years. George grew jealous over these meetings and the bonds she shared through art. More than once he accused her of having an affair. At the same time, he carried on an affair with one of his married employees, who later left the company. It was an open secret in The Woodlands offices, and Cynthia knew about it as well. She stayed in the marriage, although it added to the distance between them.[15] Cynthia stopped accompanying George to Galveston every weekend, unless it was a special occasion, and she spent more time traveling on her own to visit her grown children. They still went to dinner together during the week—every Tuesday night they ate with Cynthia's sister, Pamela—and they often attended performances together, but privately, their relationship had become strained and difficult.

Working together on the Galveston restoration project didn't heal the rifts, but it brought them together over shared interests. Unlike with The Woodlands, there was no one else to intercede. Their ideas were entirely their own. Cynthia's artistry joined forces with George's development and financial savvy. She could see what needed to be done to restore the buildings, and George had the patience to deal with the permits and city bureaucracy to make it happen. "They had a remarkable partnership in Galveston," longtime publicist Dancie Ware said. "Cynthia had the eye for detail and the design. George was the preservationist, and he was the financial side of it."

In all, George would buy more than a dozen buildings in the Strand, investing some $40 million.[16] The revitalization included turning a former mercantile building into the Tremont House, a 108-room Victorian-style luxury hotel, even though Cynthia hated Victorian style. "It's one of the worst periods in the history of design, from the standpoint of furniture and coloring and almost everything," she said. To overcome these shortcomings, they once again hired O'Neil Ford, the San Antonio architect of their first Galveston weekend home, and he designed a four-story atrium with black-and-white hand-glazed Italian tiles that bathed the interior in natural light, minimizing the claustrophobic feel of most Victorian construction.[17] George loved the hotel, where he would live in the final days of his life.

Their efforts didn't stop there. George and Cynthia acquired wharves and warehouses and turned the old port into a thriving waterfront. He bought the classic Hotel Galvez on the Seawall, a holdover from Galveston's glory days, and revived it.

George visited Galveston almost every weekend. But as he did anywhere he went, he centered his visits on business. He might take one of his sons dove hunting near the gas processing plant in Bridgeport, but they would stop and tour the plant first. In his later years, after his knees had given out and he was using an electric scooter to get around, he would roll down the streets of the Strand on Saturday morning, going from building to building and asking the tenants about their business.[18] One owner would call ahead to the next and say, "Big Daddy's on his way."[19]

George's most dramatic effort to revitalize Galveston involved the site of the old Fort Crockett, which he bought when he planned to drill the next phase of Lafitte's Gold from the Seawall. After hanging on to the property for almost a decade, he decided to develop it as high-rise luxury hotel, the San Luis Resort. A new hotel hadn't been built on the Seawall in twenty-five years, and George's vision was as grand as ever. He turned the site into a fifteen-story, 244-room resort and condominium complex rising above the old fort's bunkers. The $38 million hotel had a large pool with a swim-up bar. Palms and hibiscuses lined the entrance. It opened on June 2, 1984, amid fireworks custom-designed by New York's famous Grucci family. Writer George Plimpton, who had profiled the Gruccis in the *New Yorker*, narrated the show, telling a story about each firework before it launched. The gala was the hottest event in decades—the biggest thing to hit the island since the Maceo days. It was the kind of party that would have made Christie proud. George's phone rang incessantly from people trying to get tickets. The proceeds benefited the geriatric studies program at the island's main hospital, the University of Texas Medical Branch. "He'd never been the entertainer," said Ware, who handled the publicity. "He was the host of this great event, with a charity component, of this grand new hotel. He was so proud. He was not a proud man, but he was proud of that event."

George didn't stop there. With the San Luis open and the Strand undergoing a transformation, he decided to bring back the Mardi Gras celebrations he remembered from childhood. Like many Gulf Coast cities, Galveston had celebrated the days leading up to Lent with masked balls and parades as far back as the mid-1800s, but the festivities had died out during World War II.[20] As a boy, George would stand on the street during the parades and catch beads thrown by revelers on passing floats. It was a happy memory in a childhood marked by hardship, and he wanted to recapture that sense of wonder. He focused on the parades, because they were something that people from all walks of life could enjoy. "He used to say that the beads were democratic, because anyone who put their arms up could catch them, whether you were rich, poor, or in between, whatever color or walk of life," Ware said. "Anybody could catch a bead."

He flew to New Orleans and began negotiating with Blaine Kern, considered that city's best float maker. Kern loved to dicker, and

George never missed an opportunity to negotiate. He ordered a small fleet of floats and then convinced other Galveston business owners to sponsor them. It was classic George—getting someone else to help underwrite his dream. The Mardi Gras revival was largely his alone, yet before long, businesses across the island joined in. He loved bringing local dignitaries to Galveston—Houston mayors, famed heart surgeon Denton Cooley, and Bob Gates, the future defense secretary who was then president of Texas A&M—and watch them grab for beads along the parade route. "Look at that," he'd marvel, "just like the children."[21]

◆

The Mardi Gras revival was about as close as George came to leisure time, even though he did it in part for the good of his hometown and to increase the occupancy of the hotels he now owned there. (None of his hotels on the island would be consistent moneymakers until after his death.) People who worked with him knew that George didn't have time for frivolity. He didn't engage in small talk. Sitting next to George on a plane—in coach—meant nonstop shop talk, Colson Tough recalled. "He wasn't ordinary—let's talk about baseball or football," Tough said. "It was just business all the time."

The one exception was tennis. It was the one pastime that George cared about. "George avoided all publicity," Budd Clark once said. "He was just work, tennis, work, tennis, work, tennis."[22] (To his business associates, it may have appeared these were his only interests, because tennis cut into George's work schedule. But he spent most of his time outside of work with family or childhood friends in Galveston.) His strength as a player was ball placement rather than power.[23] He played every day from the time he entered his teens until he was eighty-three. The last five years or so, he was struggling to walk with his crumbling knees. Yet he still insisted on playing. He wrapped his knees with stretch bandages as tightly as he could and stood at the net, where he didn't have to run as much.[24]

He founded the Houston Racquet Club in 1968 because he believed the city needed a venue that focused exclusively on tennis. Too many of the Houston-area country clubs, like the one in the tony River Oaks neighborhood, were built around golf, which George didn't play despite owning numerous courses in his residential developments. He felt the golf crowd looked down on tennis, and while social cachet didn't matter to him, he wanted a club with a more relaxed, tennis-centric atmosphere.[25] He also wanted a place that would take anybody. Many of the Houston-area clubs at the time excluded Jews and minorities. George never forgot the crucial role that the Big Nine had played in financing his early years in business, and he didn't like the idea of using the wealth they helped him create to build a club that would have barred them.[26] It was one reason he never became a member of other country clubs, even ones where he frequently played. The other reason, of course, was the cost. Golf clubs had higher dues, and George didn't want to spend the money.[27] A tennis-only club made more financial sense. It had lower dues because the facilities were cheaper to maintain, and opening the club to everyone would keep the fees even lower. The more people eligible to join, the more members available to pay, reducing the cost for everyone. "He was all about a place for all people, but he was also about a place where people would pay the dues," Ware said.

George gathered a few of his regular tennis buddies and they began discussing an idea for their own club. He found some land near his Wickwood house, negotiated a bargain price, and lined up a $1.3 million loan to build it.[28] It was just a stone's throw from the Houston Country Club, where two of the city's most famous residents—future president George H. W. Bush and his future secretary of state, James A. Baker III—had met on the tennis courts in the 1950s and formed a lifelong friendship.[29] (George never played with the younger Baker, despite playing doubles with Baker's father at the River Oaks Country Club.)[30]

As the first tennis-only venue in Houston, the club touched off a wave of imitators during the 1970s.[31] It also played a key role in revolutionizing women's sports. Three years after it opened, a group of top women players met with Gladys Heldman, once a top-ranked athlete who competed at Wimbledon in 1954. After her career, Heldman published an influential magazine, *World Tennis,* and became an advocate for professionalizing women's tennis. The group included Billie Jean King, Rosie Casals, and seven others who at Heldman's urging formed their own pro tour. Heldman convinced George and other key club members to host their first tournament. She then persuaded the chair of the tobacco company Philip Morris to underwrite the competition and donate $7,500 in prize money—at the time the biggest purse ever offered for a women's-only tournament, which typically paid a fraction of what men received.[32] The fledgling women's league struggled to establish itself in the male-dominated world of professional tennis, but the effort that began with the meeting at the Houston Racquet Club evolved into the Virginia Slims tour, the cornerstone of women's professional tennis.[33]

The tour got an unexpected boost thanks to grousing from many prominent male players. The bickering culminated in the 1973 "Battle of the Sexes" between King and self-proclaimed chauvinist Bobby Riggs before thirty thousand people in Houston's Astrodome, which George attended with some of his children.[34] He didn't realize it at the time, but his insistence on an all-inclusive club laid the foundation for the establishment of women's professional tennis, which in turn opened more doors for women in other professional sports.

George himself played in amateur tournaments, and in 2007 he was inducted into the Texas Tennis Hall of Fame, whose ranks also include Bush, Baker, and the architect of George's Wickwood home, Karl Kamrath.[35] But other prominent Houston amateur players didn't feel they measured up to his formidable skills. When he was in his prime, Gerald Hines, the international real estate developer, played two sets of singles tennis a day, but he never played with George.[36] "George was a much better player," Hines said.[37] (Hines's first investment in Aspen was buying one of the Aspen Alps condos from George.)

◈

The rebirth of Galveston represented a chance for George to give back to his hometown, but it also represented something greater. It reflected his determination that business leaders should do more than simply earn profits for their shareholders—they had an obligation to tackle bigger problems.

George rarely stopped thinking about major problems confronting humankind. In the formative years of The Woodlands, he read *The Limits to Growth*, a book by three young scientists at the Massachusetts Institute of Technology that garnered worldwide attention and sold more than twenty million copies. The authors applied computer modeling (then an emerging field) and system dynamics theory (a process for understanding the behavior of complex systems) to analyze the consequences that unchecked growth of the human population posed for the planet. They studied five factors—population, food production, natural resources consumption, industrial production, and pollution—and presented their findings at the Smithsonian Institution on March 12, 1972.

The book warned that humankind's expanding ecological footprint would deplete resources to the point that mushrooming humanity would stifle global development in the twenty-first century and diminish the quality of life.[38] Consumption of natural resources, food, and energy was growing at an unsustainable rate, as were the side effects like pollution. The book claimed that the cost of producing the goods needed to live would rise dramatically, and societies would devote a growing percentage of their gross domestic products to such necessities as energy and clean water. Social decay would ensue within decades. The only solution was embracing technological changes and social values that curbed unchecked growth. Adopting recycling, curtailing pollution, and focusing on food production and services rather than industrial output would help mitigate the decline.

The premise resonated with both George and Cynthia, who tended to influence his thinking on nature and the environment.[39] "I read the book . . . and it just impressed the hell out of me," George said. "The concept was important."[40] However, he took issue with the conclusion that government should intervene to limit growth. How, for example, could countries reasonably put a limit on their populations? How could developing countries be denied resources to improve their living standards? "I knew there had to be limits on some growth, but not limits on other growth," he said. Growth that depleted too many resources or produced too much pollution was bad, but growth that used resources efficiently and improved living standards was good. George argued that economic expansion was essential to funding sustainability, a belief that put him decisively at odds with many environmentalists. And his concerns about pollution, sustainability, and the environment put him in opposition with his fellow oilmen.

Despite what he saw as its flaws, George believed *The Limits to Growth* was onto something. As he often did when he read something he found interesting, he picked up the phone and called the lead author, in this case Dennis L. Meadows at MIT. After several discussions, George offered to host a series of conferences in The Woodlands to discuss the ideas in the book. The sessions often grew lively, because George invited scholars with opposing views, including Herman Kahn, a founding director of the Hudson Institute, who argued that humanity was on the cusp of a golden age of prosperity—essentially the opposite of Meadows's predictions.

"What we were really seeking is the nature of sustainable societies," George said. "I decided that should be the theme of the conference. I think that was the first time the word "sustainable" had been used in terms of growth studies."[41] More than three hundred experts from around the world attended the first three-day conference, which the irreverent Texas newspaper columnist Molly Ivins dubbed the "Doom Conference." After sitting through the first day of the academic and often esoteric discussions, she wrote: "I have seen the future, and it is polysyllabic."[42]

Meadows suggested that George sponsor an international competition for scholars, ecologists, and other scientists who offered the most creative solutions to managing large-scale growth.[43] George and Cynthia loved the idea. Known as the Mitchell Prize, $100,000 was awarded to the winner at The Woodlands conferences. "Our cities are in deep trouble all over the country," George said. "The middle class has fled; the brain base is gone; it'll take 60 years, three generations, to stop the deterioration and break the poverty cycle. We have to do *something*. That's one of the reasons Cynthia and I started offering the prize."[44]

In 1984, George went a step further. He created *The Woodlands Forum*, a quarterly newsletter produced by his public relations staff, which sought out interview subjects who often disagreed with him. George wanted a publication that would present all sides of social and environmental issues in hopes of getting the attention of politicians and other business leaders. Among the interview subjects were author Isaac Asimov, media mogul Ted Turner, physicist Amory Lovins, and Tim Mahoney, the Sierra Club's Washington lobbyist.[45] "We were allowed to do our own thing as journalists, pretty much completely," said Tony Lentini, who worked on the project.

The Woodlands conferences were the first attempt in the United States to address growth concerns. Whether they were successful depends on how you define the term. Initially, the gathering got a lot of media attention. George brought in directors with a national reputation, and big names such as former president Jimmy Carter, whom George had backed when Carter ran for office and who had become a friend. Although George thought Carter was "too religious for his own good," he respected the former president's humanitarian work and his commitment to improving societies worldwide. "I have a lot of admiration for President Carter," George said in 1988. "He's smart. There's no doubt that people malign him a little bit about the problems that were in his administration, but he is really talented, and I think he's going to do some good."[46] But the conferences had little follow-up on the discussions. By the mid-1980s, Texas had fallen in the grips of a crippling oil bust, and the state wanted to diversify its economy away from a singular dependence on the oil and gas industry. It saw technology as the next economic boom, and George did too. Technology was one of author Meadows's tools for curbing the problems of growth, and for George it became a big focus.

About the same time, and without any conscious connection to The Woodlands conferences, engineers within Mitchell Energy were also embracing new technology to solve a major problem. It would make a greater contribution to reducing pollution and energy costs than most of Meadows's ideas, but by the time it was developed, no one outside the oil industry would celebrate it as a sign of progress.

CHAPTER 17

Subatomic Dreams

"What can I do for you?" Peter McIntyre hesitated for moment, caught off guard by the voice on the other end of the phone. A high-energy physicist from Texas A&M, he had called The Woodlands office of Mitchell Energy and Development to find out how he might make a proposal to George Mitchell. He'd just reached George's secretary, who had put him on hold. He'd expected her to come back on the line and take a message. Instead, he found himself on the phone with George.

McIntyre stammered for a moment. What he wanted wasn't easy to explain. Working with other physicists, he hoped to build a fifty-four-mile circular underground tunnel that would raise the world-wide standards for high-energy physics research. McIntyre and his colleagues wanted to shoot beams of subatomic particles known as protons in opposite directions around the tunnel, steered by giant electromagnets, and then slam them into each other. They hoped that the collision would produce a long-sought particle, the Higgs boson.

As he spoke, McIntyre couldn't help but marvel at the absurdity of the situation. He had never talked to someone as wealthy as George Mitchell, and he didn't know the protocol for asking for millions of dollars. He rambled on about quarks, superconducting magnets, and the huge underground tunnel in Texas, speaking faster as he went through his explanation. He was sure George would hang up on him at any moment.

Until a few days earlier, he had never even heard of George Mitchell. McIntyre had been working with a group of colleagues who were trying to improve the sorry state of high-energy physics research in the early 1980s. America had dominated the effort to understand the building blocks of matter since 1934, when physicist E. O. Lawrence invented the cyclotron, the first circular particle accelerator, which revealed radioactive isotopes and smaller subatomic particles that were valuable for medical and scientific research. After the development of the atomic bomb during World War II, US physicists focused on finding the fundamental elements of matter. In the postwar years, the quest was fueled by Cold War tensions and a link between particle physics and national security. The United States built ever-larger particle accelerators, which blasted beams of subatomic particles into each other at increasing speeds. The collisions led to the discovery of tinier particles such as quarks, hadrons, taus, and neutrinos. Colliders at the Brookhaven National Laboratory on Long Island; the Fermi National Accelerator Laboratory, or Fermilab, in Illinois; and the Stanford Linear Accelerator Center in California all contributed to American leadership in high-energy physics.

The Arab oil embargo of 1973 changed America's energy priorities. The Carter administration rolled high-energy physics research into the new Department of Energy, where it competed for funding with solar power and alternative fuel programs that had more support. Government money grew more difficult to obtain as the accelerators became bigger and more expensive. European scientists began to pull ahead.

To maintain America's lead, US physicists focused on superconductivity—the concept that at temperatures near absolute zero, electrical current flows through certain metals and alloys with almost no resistance. Magnets wound with coils made of these alloys can generate massive magnetic fields with far less power consumption. Superconductivity would allow particle accelerators to achieve even greater speeds, potentially unlocking new discoveries, with far less power consumption—and less cost. "Superconductivity is a magic potion, an elixir to rejuvenate old accelerators and open new vistas for the future," physicist Robert R. Wilson, a Manhattan Project veteran and the architect of Fermilab, wrote in *Physics Today* in 1977.[1] With essentially no resistance to the flow of electricity, particle accelerators could double their flow of electrical current while saving millions of dollars on their electric bills, and it could all be built at "modest cost." The electricity savings were important because colliders used a lot of power.

None of the existing labs had magnets big enough to accelerate particles to the higher velocities physicists needed to find new particles.[2] They envisioned a new collider that would be larger and more powerful than any previously built. Funding, though, continued to be a problem. In 1983, President Ronald Reagan, working to reduce the federal budget deficit, ordered a freeze on big new government projects. To fund his tax cuts, Reagan slashed government spending, and energy research projects were particularly vulnerable.[3] A new collider, despite the "modest cost" of superconductivity, would still require billions of dollars.

McIntyre had been among the physicists working on the early proposals for this Superconducting Super Collider, and although the Department of Energy did provide enough funding for initial research, getting the money to keep it going would require some proof that the collider could work.

The project needed a lot of land—the collider ring would be an underground loop roughly the same distance around as the Washington Beltway—and while some scientists had proposed other locations, McIntyre knew that Texas had something few other places could offer: wide open spaces close to a major airport. He wanted to build the collider just south of Dallas, near the town of Waxahachie. The site had something else going for it: geology. Waxahachie sits atop the Austin Chalk, a Cretaceous geological formation that follows the Gulf Coast from Mississippi to Mexico. The chalky soil was easy to dig yet stable enough that it wouldn't cave in.[4]

McIntyre needed to demonstrate that the technology was cost effective, and that meant building prototypes of the giant magnets that would steer the light beams around the tunnel. Those prototypes would cost millions of dollars. Unlike research in medicine or computer technology, innovation in high-energy physics depends almost entirely on government funding. In the austerity of the Reagan years, the DOE made it clear it wasn't interested in funding a multibillion-dollar collider project. Besides, the department had just funded an upgrade of superconducting magnets at Fermilab's Tevatron.[5]

Frustrated and unsure where he would get the money for the prototypes, McIntyre had been lamenting the situation to his secretary when another longtime physics professor, Tom Adair, walked into the office. Adair told McIntyre that he should call George Mitchell. Adair had never met George, but he knew that George had recently set up the Houston Area Research Center in The Woodlands. As George's interest in saving the planet by curbing "bad" growth had shifted more toward technology, he'd established HARC as a nonprofit organization in which four major Texas universities—A&M, UT, Rice, and the University of Houston—could conduct cooperative research and share projects that were too big or too ambitious for any single institution.[6] He raised $90 million and distributed it through venture capital funds with specific goals, such as biomedical research.[7]

Originally, George developed HARC to attract more professional jobs to The Woodlands, but he soon saw that much of the research dovetailed with sustainability.[8] "He wanted to get the major universities together to do applied science," George's son Scott said. "He saw a short circuit between business and science. Why aren't business and science more aligned? He saw a lot of good science going on that wasn't being used. And why isn't the science community focused on business applications and business challenges? He wanted business to have the advantage of science. That was the rationale behind HARC." Superconductivity, for example, wouldn't just make it cheaper to slam protons together; it could also make it cheaper to generate and transmit electricity for consumers.[9]

Adair knew that George was funding HARC's A&M faculty member, and he knew George was an "Old Ag" who had donated money to the university. That's who you need to call, he told McIntyre. He's one of the few people in Texas who is both a wealthy and successful businessman and who also has enormous vision and an interest in science.

This was how McIntyre wound up stumbling through his pitch in one of the more outlandish cold calls in the history of the telephone. "I figured at any moment he was going to say 'I've got a meeting' and pass me off to someone else," McIntyre said. "But the next thing I know he says, 'I suppose you ought to come down and meet with me.'"

Two days later, McIntyre made the hour-and-a-half drive from College Station to The Woodlands, and this time he was better prepared, bringing drawings, physics papers, and computer printouts. George listened intently as McIntyre explained the collider and physicists' desire to confirm the existence of the Higgs boson—at the time the holy grail of particle physics. McIntyre tried to strike a balance with his explanation. He needed George to understand the importance of the project, but he knew George wasn't a physicist. He didn't want to overwhelm him.

McIntyre explained that for most of the twentieth century, physicists had gradually expanded their understanding of the basic building blocks of matter—starting with atoms, then protons, electrons, neutrons, and later, particles such as quarks. They also knew, however, that the universe comprised more than just matter. In the 1970s, particle physicists developed the Standard Model, which postulated that the universe is made up of twelve particles of matter and four forces. The Higgs boson was the basic element of a field that gives particles such as quarks their mass, and scientists believed that mass slowed particles down after the Big Bang and led to the creation of atoms. Because it's the fundamental element of the universe's other building blocks, the Higgs is sometimes called the "God particle." The Standard Model told scientists that the Higgs boson existed, but they couldn't prove it. The Super Collider would accelerate protons to unprecedented levels of energy traveling clockwise around the tunnel. Then, scientists would slam them with another proton beam traveling counterclockwise at the same speed. The resulting collision would shatter the protons, creating new subatomic particles, at least for an instant. Researchers would monitor the explosion with sensors that they hoped, in that instant, would reveal the Higgs boson.[10]

McIntyre finished his explanation, and to his surprise, George seemed to be following him. "Probably three hours went by," McIntyre said. "We were down on the floor of his office, with the blueprints spread out of the superconducting magnet designs. Finally, George said 'tell me about the politics.'" McIntyre told

George about the reaction physicists had gotten from the DOE.

Mitchell thought for a moment and then said, “You’re going to need to set up a lab.”

Yes.

“You’re going to need some money for that.”

Yes.

“How much?”

McIntyre gulped. He was a physicist, a college professor, a research lab rat. He’d never asked anyone for millions of dollars. Sheepishly, he told George he was still trying to figure out the exact amount. It was only then that George flashed a sign of impatience.

“When you get to the stage you’re at with this thing, you need to do your homework and figure out the right answers when somebody asks you a question like that,” he snapped. “Because if you’re low-balling, they’re going to take you up on it and you’re going to make a mess of things, and if you high-ball it, they’re going to tell you to go to hell. Now, how much money do you need?”

Surprised by the rebuke, McIntyre said he needed $3 million. With the number on the table, George lapsed into his negotiating mode. He loved haggling over deals, even if it was with a novice who had come to him for help. Times were tight, he said—it was the mid-1980s, and oil prices were falling, taking natural gas with them—but he would put up half the money. George would use his contacts at A&M to convince the university to kick in the other half. But, he told McIntyre, they also needed the government to match what they raised. McIntyre said that might be tough, given the budget constraints in Washington. George said he would make a few calls on that, too.

McIntyre didn’t know it, but it was the same funding model George had used to pay for Mardi Gras floats in Galveston, and for that matter, most of his early business ventures—get others to share in the financing. When he gave to charities, he put up part of the money and encouraged the charity to use his donation to match the remainder from other sources. His purpose was twofold. George believed that a worthwhile cause should have broad support, and the matching-fund model challenged others to contribute as well. In some cases, it even created a sense of competition among donors.

In the end, the tactic often raised more money than if he had written the entire check himself.[11]

As they put the funding together, George offered McIntyre lab space at HARC. Over the next few years, a couple dozen physicists, engineers, and other top scientists, working with counterparts at universities and research facilities around the country, developed three magnets, each thirty-five meters long. These were critical to the collider's design because the magnetic field they generated kept the proton beam in place as it traveled around the collider's arc. George demanded the same sort of progress reports from McIntyre as he did from his petroleum engineers.

"Every few weeks, he would want us to come over and brief him on what we were doing," McIntyre said. At one of the meetings, McIntyre walked in to find George huddled with a team of geologists studying seismic maps. He asked George what he was working on, but George waved him off. "It's just an idea I've been playing with for a while now," George said. "It's no big deal. Come tell me about your magnets." McIntyre didn't think much of it. After all, he wasn't a geologist, but he assumed they were always working on new ways for getting more natural gas out of the ground.

⬥

Over the next four years, the design for the Superconducting Super Collider came together, and McIntyre's choice for the site south of Dallas was embraced by other teams of physicists around the country, as well as the Department of Energy.

The ring would make a fifty-four-mile loop, starting just north of Waxahachie and swinging south almost to the Ellis County line. Ringed with 4,728 magnets, each seventeen meters long, it would have twenty times the collision energy of any previous collider. The magnets alone required 41,500 tons of iron, the equivalent of four Eiffel Towers.[12] The project would have an independent power grid that could routinely deliver ten megawatts of electricity—enough to serve about ten thousand homes.[13]

By 1987, the initial work of McIntyre and the other teams of physicists showed that the Super Collider was feasible, if expensive:

construction would cost an estimated $4.4 billion. Despite the budget concerns, the project had its champions within the DOE, and they briefed Reagan on the initiative. They needed his support to overcome congressional reluctance to new spending. Reagan listened to the briefing, and when it was over, he pulled an index card from his suit pocket. On it was "Credo," Jack London's poem about seizing opportunity. He explained that Oakland Raiders quarterback Kenny Stabler once told him the poem means "throw deep." And with that, Reagan approved the funding. The president's support ensured the project would move forward.[14] There was even talk of calling the collider the "Gippertron."

In the early 1990s, crews carved out about fifteen miles of tunnels, with seventeen access shafts, and built more than two hundred thousand square feet of office and lab space. The project directly employed about two thousand people, including some two hundred scientists, and it promised to create as many as thirteen thousand more jobs as the collider reached fruition.

Almost from the time Reagan approved it, however, the Super Collider caused conflict between the physicists and the accountants at the Department of Energy. To this day, McIntyre claims that the DOE's mismanagement doomed the project because the agency favored a more expensive and less effective magnet design than the one he wanted.

"I would keep George abreast of this stuff," McIntyre said. "It wasn't always what he wanted to hear, but he always wanted to hear the truth." The truth was that by the fall of 1993, the Superconducting Super Collider was in big trouble. The project's original $4.4 billion price tag had ballooned to $11 billion because of magnet redesigns, spiraling administrative costs, and expenses such as land acquisition that hadn't been included in the original estimates. The plan called for attracting contributions from foreign governments to help cover construction costs, but none of the seven countries that the DOE hoped to bring in ponied up any money.

Members of Congress, battling budget deficits and the lingering effects of recession, concluded that the Super Collider was a costly mistake and voted to cancel it.

George was furious. Just as he chafed at business leaders who were too shortsighted to look beyond the next earnings report, he believed politicians had killed an important project for political expediency. He worked his connections, but the Texas delegation already supported the collider. It probably didn't help that he had backed fellow Greek American Michael Dukakis in the 1988 presidential election over Houstonian George Bush, whom George had known since they teamed up against bayou channelization in the 1960s. George's politics could be difficult to interpret because he wasn't beholden to party affiliation. He considered himself a "moderate independent" and said, "I'm more Republican than Democrat, but I work with liberals, too." He would later praise Bush for his handling of the Persian Gulf War, but he frequently criticized both Bush and his predecessor, Reagan, for what he saw as ineffective energy policies. George wanted the government to impose tariffs on oil imports that would set a floor of twenty-five dollars a barrel on prices and encourage domestic drilling, and he wanted Bush to promote natural gas. He felt that both Bush and Reagan had neglected energy conservation.[15]

The Super Collider vote was just one more political disappointment for him. He believed that Congress had ceded America's longstanding leadership in Big Science to the Europeans. If the project had been completed, it would have had five times the energy of the Large Hadron Collider in Europe that ultimately discovered the Higgs boson in 2012. Many physicists believe that the Super Collider would have detected the Higgs boson years earlier and reasserted American dominance in particle physics.[16]

George's anger would simmer during the 1990s and inadvertently lead to a lasting friendship with the world's most famous living physicist. In the meantime, though, he had bigger frustrations brewing at Mitchell Energy.

CHAPTER 18

The Fight for Fracking

By the late 1980s, Mitchell Energy and Development faced deteriorating long-term financial prospects. The company had weathered the boom-bust cycle of oil early in the decade—one of the most volatile periods in the industry's history. Interest rates soared, increasing the cost of the company's debt, as oil and natural gas prices fell, which meant it had less cash coming in to pay the higher financing costs. George had been struggling with lenders almost since the company went public in 1972. Maintaining gas production in the Boonsville Bend still required huge amounts of capital, as did the company's other drilling prospects. Now, Mitchell Energy had the added costs associated with The Woodlands. George and other executives bristled at the growing constraints from the company's primary lender, Chase Manhattan. When George decided to change banks, David Rockefeller, the bank's chair and scion of the country's first oil magnate, flew to Houston to change George's mind, without success.[1]

Still, there weren't a lot of options. By 1986, six of the state's biggest banks were on the hook for $11 billion in outstanding energy loans.[2] As prices slid below ten dollars a barrel, many banks refused to lend more money to the industry. Just as the Big Nine had found Mitchell Energy's capital needs more than they could keep up with, banks questioned the company's insatiable thirst for capital and its ability to maintain growth.[3]

George managed to make it through the 1980s without losing money, and he was able to keep the banks at bay, but Mitchell Energy didn't emerge unscathed. The company had about 3,500 employees in 1983—in both energy and real estate—but by 1988, it had cut its ranks to 2,200 through early retirements, attrition, and layoffs. "It's a tough thing to cut back that much," George said. "But we did; we had to. We had to make it work."[4] As difficult as it was, Mitchell Energy fared better than many other independents, who either were acquired or filed bankruptcy.

Since the company went public, George had tried convincing investors, Wall Street analysts, and anyone who would listen that owning oil and gas and real estate made sense, but few could see the benefits. What's more, while some oil companies had ventured into real estate as a hedge against oil price fluctuations, Mitchell Energy had diversified in the same geographic region.[5] Its biggest gas field was in North Texas, and most of its real estate was around Houston. Because of the prominence of the oil industry in the Texas economy, real estate values were closely linked to oil prices. That meant that when oil and gas fell, both sides of the business suffered. What's more, George often said that The Woodlands would take fifty years to pay off, far longer than most investors would wait for a return. Even within the company, it was becoming clear that the analysts had been right all along: for the good of the shareholders, the company needed to split in two.

Despite these long-term concerns, Mitchell Energy reported strong financial results from its oil and gas operations. In 1991, it earned more than it had during the previous five years of the oil slump. Iraq's invasion of Kuwait drove up prices, pulling land values in Houston skyward along with them. Mitchell Energy had become one of the country's biggest producers of natural gas liquids, thanks to the processing plant in North Texas that hummed along. George personally was worth about $500 million, counting his 63 percent stake in Mitchell Energy and his private investments. But Mitchell Energy's stock value depended almost entirely on the energy side of the business. Its real estate holdings had negative cash flow, and George didn't expect that would reverse anytime soon. In other words, the company was taking profits from the gas business and using them to cover losses in real estate. Despite George's efforts to lure good-paying jobs to The Woodlands, the development still lacked enough office space. Houston was working through a commercial building glut from the 1980s, when developers overbuilt, and the city became known for its "see-through" buildings. The excess put a cap on lease rates, which hurt George's development plans. In The Woodlands, offices rented for about fourteen dollars a square foot, but the company needed seventeen dollars to build a six-story building profitably.

Wall Street began pressuring the company to spin off its property business, so it could plow more profits into expanding its energy operations. George stubbornly refused to address the issue. "I haven't decided what we should do, if anything," he said, when asked about splitting the company.[6] Because he controlled two-thirds of the stock, he didn't worry too much about mollifying investors. Besides, not all institutional shareholders were scared off by George's long-term strategy. Some liked the fact that he wasn't beholden to meeting Wall Street's quarterly earnings projections. "We had far more institutional investors who opted out rather than opted in, but the ones who opted in were buying into that vision," said George's son Todd, who joined the Mitchell Energy board in the 1990s.

While George remained committed to The Woodlands, he also couldn't deny he was getting closer to retirement. In 1976, when he was fifty-seven, George had hired Don Covey, a twenty-one-year Shell Oil veteran, and four years later had named him company president and head of the energy business. Many employees assumed Covey would take over the company someday, although George didn't designate an official successor. As late as 1991, he told *Forbes* he planned to keep working "for at least the next two or three years."[7]

Covey was well liked, and he possessed a personality and management style similar to George's. It was Covey, a deeply religious man who seemed to genuinely care about people, who helped tone down the notorious shouting matches among upper management, especially between George and Budd Clark. George's temper flared and receded quickly. "He would get in a hell of a knock-down, drag-out argument with you, but when it was over, it was over. The next day, he'd be smiling at you," said Homer Hershey, the company's longtime senior vice president of production. Not everyone could handle being chewed out by the boss one day and patted on the back by him the next. Covey brought an even-keeled temperament that created a buffer between George's flare-ups and his employees.

"Covey was unique in that he could handle George," Hershey said.

In November 1993, Covey, who was fifty-nine years old, was en route to London with his wife when he collapsed in midflight from an aortic aneurysm, a rupture in the main artery from the heart. The pilot, determining the aircraft was closer to London than to New York, continued to Heathrow Airport, but by the time the plane landed, Covey was dead. His sudden passing at such a young age hit the company hard, derailing any succession plans that George may have had.[8] "That took the wind out of George's sails when Don died," recalled Herb Magley, a geoscience manager at the company for twenty-six years. "It changed the whole attitude." George had an empathetic relationship with many of the company's employees, but he was particularly fond of Covey. "He just thought the world of him as person," George's daughter Sheridan Lorenz said.

George was still chair and chief executive officer, but he was seventy-four by then, and much of the board and senior management was getting older as well. Budd Clark was seventy-two, and most other senior executives were in their late fifties.[9] The company needed new blood. George's sons Todd and Kent were directors, but Todd had started his own energy company and had little interest in taking over the family business. Kent was involved in George's private real estate interests, primarily the Bald Head Island beach community in North Carolina. George's other eight children all had their own careers, and although some overlapped with their father's—his youngest son, Grant, for example, was involved in the family investments and the Galveston projects, and his oldest son, Scott, was a developer who had built homes in The Woodlands—none was in a position to take over the company.

While he stubbornly refused to issue shares to boost the company's liquidity, George grew increasingly frustrated that Mitchell Energy's stock wasn't worth more. It started the 1990s at almost twenty-two dollars a share, but by early 1992 it had dipped below fourteen dollars.[10] "He couldn't understand why they couldn't get the stock price up," publicist Dancie Ware said. "He always thought the company was grossly undervalued by the analysts."

George wasn't the first CEO who thought Wall Street didn't understand his business, but he also wanted the company to be more profitable. Mitchell Energy was big enough that it needed to shed some of its familial ways. Once again, George was looking for an executive to instill more discipline and professional management.[11] But he also wanted someone who understood the company. He found the solution sitting on his own board.

◈

In 1992, W. D. "Bill" Stevens got a phone call from George suggesting that they have lunch. Stevens had spent the previous thirty-five years at Exxon Corp., the last four as president of its US operations. A year into his presidency, the tanker *Exxon Valdez* ran aground on Bligh Reef in Alaska's Prince William Sound. The accident ripped open the tanker's hull, spewing eleven million gallons of crude oil into the environmentally sensitive area and despoiling some 1,300 miles of coastline. The disaster forever tainted the image not just of Exxon, but of the entire oil industry. Stevens flew to Alaska during the first two days of the spill and took much of the blame for the catastrophe.

He had been popular with many Exxon employees, but he often clashed with senior management. After the *Valdez* incident, he opposed job cuts and other cost reductions ordered by top executives, and while he left voluntarily, he left under a cloud.[12]

Stevens didn't know George well, and the few times they met, they typically found themselves on opposite sides of the issues. Stevens, after all, represented the biggest of the major oil companies, and George had long been a voice of the independents. "We never had any real altercations, but then we never agreed on anything either," Stevens later recalled. If he was surprised by George's call, he was even more surprised when he got to the lunch and heard what George had to say. He wanted Stevens to join Mitchell Energy's board. "Are you sure you have the right guy?" Stevens asked in disbelief. George said despite their past differences, Stevens was exactly who he wanted as a director.[13]

Stevens had experience running a large oil company and the deep understanding of financial discipline that Mitchell Energy desperately needed. With his appointment, the tone of the board changed almost immediately. Stevens's years at Exxon had taught him to look ahead and spot potential trouble. George, of course, thought frequently of the future, but not in the same way. For one thing, he tended to look decades down the road, and his visions were grand and sweeping. He didn't concern himself with how his ideas would be achieved. Stevens worried about where problems might crop up in two or three years. One area that concerned him in particular was the contract to supply gas to NGPL from Mitchell Energy's properties in North Texas.

The agreement had been extended several times, thanks to George's negotiating skills. The last time he had convinced NGPL to renew its agreement, he had been in Austin for his granddaughter's birthday party. George entered the house to find it filled with screaming five-year-olds. He asked his son-in-law whether there was quiet place he could make a call. The quiet room was an upstairs bathroom, and George sat on the toilet and renegotiated a deal that would keep his company operating for years to come. Then he calmly went downstairs, walked up to his son-in-law, and asked, "Is there any cake and ice cream left?"[14]

Despite George's skill as a negotiator, though, he and his executives knew that sooner or later, NGPL would balk at paying above-market prices for gas. By the 1990s, most energy companies believed oil and gas prices would remain cheap for a long time. Mitchell Energy had been supplying 10 percent of Chicago's gas for decades, and NGPL was paying about five dollars per thousand cubic feet. The market price was less than one dollar.[15] When the contract expired, even if George renegotiated it, Stevens warned, Mitchell Energy could expect to receive less for its gas. And that was a serious threat to the company's financial health because the NGPL contract accounted for about one-quarter of Mitchell Energy's total production. Once that deal ended, the company would never get anything similar. Like many Wall Street analysts, Stevens worried that the company's dual structure was holding it back. In the twenty-year debate internally and externally about how to handle the disparate businesses, Stevens favored focusing on energy. He could see what George chose to ignore: the duality wasn't working. Real estate was a drag on the energy company, and that's why the stock price was suffering. It didn't matter if The Woodlands paid off in fifty years or just twenty. If the company wanted to survive the next ten, it needed to split in two. The challenge was convincing George to accept the idea.

Stevens was a lone voice among a dozen directors, many of whom were fiercely loyal to George—two, after all, were his own flesh and blood. But after Covey's death, Stevens emerged as George's choice to succeed him as company president. It made sense, at least practically. Stevens was already on the board and knew Mitchell Energy. Like Covey, he had extensive experience at a major oil company, and he was only fifty-nine years old—the same age as Covey had been—which meant he could continue to guide the firm as George eased toward retirement. While George made the decision, other board members saw the appointment as a positive. The *Wall Street Journal* even hinted that Stevens's appointment could mean that Mitchell Energy was girding for a major expansion.[16]

From the start, Stevens was a tough sell with many of the Mitchell Energy faithful.[17] To them, the company was as an extended family, and for many employees, losing Covey felt almost like losing a relative. Perhaps anyone stepping into the role would have faced skepticism, but Stevens, despite serving on the board for a couple of years, was an outsider. Plus, his gruff personality clashed with Mitchell Energy's congenial atmosphere. "He's not your warm fuzzy friendly type," said one former executive who worked with Stevens at Exxon. Stevens tried to rekindle the same good will he had garnered among Exxon employees. His first day on the job, he stopped at every desk and introduced himself to all the employees, asking their name and what they did at the company.[18] Many employees, however, remained wary. George showed occasional frustration with Stevens and once commented that Stevens could be hardheaded, but he also recognized he needed someone who would make tough calls. George was too emotionally involved in his businesses, especially The Woodlands, to make those moves. "The company badly needed someone who was really strong and willing to make decisions that were unpopular," Todd Mitchell recalled. "[Stevens] didn't care what people thought about him. He didn't care if he hurt people's feelings."

Soon after Stevens took over, heavy rains flooded part of The Woodlands. Mitchell Energy's public relations man, Tony Lentini, suggested that they take a group of about twenty company volunteers and help the residents clean up their homes. Stevens seemed puzzled by the idea. Lentini said it would foster good relations. Besides, many company employees lived in The Woodlands and wanted to help their neighbors. Stevens said he had never done that kind of thing at Exxon, but he grudgingly allowed Lentini to take five company volunteers.[19]

While employees may have bristled at Stevens, George and the board gave him more power than Covey ever had. While Covey had simply been president, Stevens gained that title as well as the newly created job of chief operating officer. The new COO position meant Stevens was now essentially running the day-to-day operations, in both energy and real estate.

Even outside the company, people noticed the difference in style between George and Stevens. George had long been unusual among oilmen in that he had few qualms about talking with the press. While he didn't socialize much within the industry elite, he was well known in Houston, and despite their best efforts to avoid publicity, he and Cynthia often landed on the society pages for their philanthropic activities. When it came to company business, George was equally open with reporters. Stevens, on the other hand, rarely gave interviews, perhaps having grown wary in dealing with the aftermath of the *Valdez* disaster at Exxon.[20]

"He wasn't an entrepreneur, and he didn't have that spirit of an entrepreneur," Dancie Ware said. "He was much more of a CEO, [and] they needed that CEO management to move ahead."

Stevens brought a strong sense of financial discipline to the job. As the decade progressed, both he and George grew increasingly concerned about the company's long-term health. In the near term, Mitchell Energy remained relatively strong. It was replacing the oil and gas it produced with new reserves, a measure of its sustainability, but Stevens saw that as the decade wore on, replacing reserves would become increasingly difficult. The company's problems were much greater than just addressing the end of the NGPL contract. For publicly traded companies, reserve replacement is a financial treadmill. If the total reserves fall, so does the company's stock price. But maintaining reserves requires additional capital for new drilling, and Mitchell Energy had no new plays like the Boonsville Bend to save the day. The gas fields in North Texas were still prolific, but the high costs to keep them producing meant that the Boonsville Bend wasn't the profit center it had once been. Because of its heavy debt burden, Mitchell Energy couldn't easily borrow more money, and George was as steadfast as ever that he wouldn't issue new shares to boost liquidity. How the company should resolve this conundrum created a rift in management's top ranks that pitted Stevens's resolve against George's vision for the future. George saw the company's salvation in the experiment that Peter McIntyre, the Texas A&M physicist involved in the Super Collider, had spied geologists working on in a corner of the Mitchell Energy office back in the early 1980s. For Stevens, that idea was pure folly.

◈

George had worried about the company's crown jewel, Boonsville Bend, as far back as the late 1970s. For twenty-five years, Mitchell Energy had steadily produced gas from the three hundred thousand acres it held in North Texas. But by then, some of the gas processing equipment needed an upgrade. NGPL offered to pay for the improvements if George promised that the former Frustration Fields would keep producing gas for at least another five years. George gathered his geologists and engineers and told them to conduct a study, but rather than limit the outlook to five years, he wanted them to estimate how much longer Boonsville Bend would last before production declined. Their analysis found that the gas would keep flowing for about a decade more. But it also showed that by 1992, Mitchell Energy would have trouble drilling enough new wells to replace what it produced, and its reserves would begin to decline. Ultimately, Mitchell Energy would fail to meet the production volumes required under the NGPL contract. In other words, in about a decade, the company's North Texas ATM would stop spitting out cash.[21] This decline was exactly what Bill Stevens saw when he took over as CEO.

George didn't know how to solve the problem, but he reminded his geologists and engineers that the company's future depended on finding an answer. They already had an idea of where to look. Oil and gas fields aren't defined just by square miles on the surface; they're also defined by depth. The Boonsville Bend—geologically, it was known as the Atoka conglomerate—ran under the North Texas hills at a depth of about 3,500 feet. Below that, there were other geologic layers, some of which might also hold gas. Beginning at a depth of about 6,000 feet, there was a layer of dense, slatelike rock that geologists called the Barnett Shale.

In 1976, the Department of Energy funded the Eastern Gas Shales Project in Appalachia, which demonstrated that natural gas could be extracted from shale formations in the region. If gas could be recovered from those shales, maybe it could be drawn out of the Barnett, too. That might be just the solution the company needed, because George already had the mineral leases to drill there. But there were a few problems. Not all shale formations are the same, and the ones in Appalachia differed from the Barnett. The findings of the DOE study might not translate for North Texas. Besides, the report concluded that while the gas was in the shale, drilling into it didn't produce enough to make it worth the expense.

Then, in 1981, a paper from one of George's geologists showed up on his desk. Just as he reviewed the data on each well the company drilled and fretted over street signs and seemingly every tree cut in the Woodlands, he also insisted on reviewing every paper from his employees before they submitted them to industry trade journals. The paper, by Jim Henry, indicated that the Barnett Shale was the source of the gas in the Boonsville Bend. The Barnett, Henry concluded, had large natural fracture systems that might make it possible to extract the gas. He wasn't suggesting that companies should drill into the shale—the process wasn't economically viable—but that getting the gas out would be possible by widening natural fractures. The more George read, the more excited he became. If the Barnett Shale held more gas, it would be as if the company had discovered another field under the one it already had. That would be a lot cheaper than finding a new gas field somewhere else.

George wanted to learn what secrets the Barnett Shale held. In

1981, with the skyscrapers of downtown Fort Worth on the distant horizon to the southeast, Mitchell Energy drilled a well named the C. W. Slay #1. The results were unimpressive. This was no second coming of Spindletop, and George would not be dancing around the well like Jett Rink in *Giant*. Then again, George wasn't looking for a eureka moment. He knew the project would take time and careful study. The Hollywood version of oil-field success was decades out of date. In reality, a new discovery was far less dramatic. There was no gusher or visible sign of striking oil. Instead, the well was hooked up to a pipeline and engineers studied the flow of oil and gas for months to determine whether it would produce consistently in commercial quantities. The process was dull and tedious.

But as George's geologists studied the results from the C. W. Slay, they found the first hint of what the Barnett might offer. "We noticed that when we drilled through the Barnett, it was like drilling through a snow bank," said John Hibbeler, the senior vice president for exploration. "We got thousands of units of gas coming out."

The presence of so much gas was intriguing but didn't solve Mitchell Energy's problem. Shale was as solid as a tombstone, and every first-year geologist knew that it wasn't porous enough to produce oil and gas. "It was terrible looking rock, but it had a lot of gas in it," Hibbeler said.

What geologists *did* know about shale was that it was source rock, meaning it was the origin of the gas that Mitchell Energy had been producing from the shallower formation of the Boonsville Bend for thirty years. Shales like the Barnett are prevalent across the middle part of the country, in the giant bowl of a long-dead ocean that once stretched between the Rocky and Appalachian mountain ranges. As plants, plankton, and other marine organisms of this long-ago ocean died, they became pressed into the sediment and clay on the seabed. Layers built up over time, and the dead organisms baked in the heat and pressure of the earth. Over millions of years, the decaying organic material turned into hydrocarbons—oil and gas—which migrated from shales, the hardest of the compressed sediment, to more porous rock like limestone.

The main difference between oil and natural gas is the size of the molecules—gas molecules are smaller.[22] In the twentieth century, when humans began to drill into the rock, it released the oil and gas that had collected in the pores of the limestone. The shales, though, had no such pores. They were as dense as a blackboard. Back when George attended Texas A&M, he had learned that shales did not give up oil and gas except through natural fractures, and nothing in the ensuing forty years had changed that line of thinking.

But George didn't want to hear about the limitations of conventional wisdom. If he could ponder problems like overpopulation and sustainable development, as a veteran oilman he ought to be able to find a way to tap into source rock and extract natural gas. At one point, he told his shale team: "If you don't think you're capable, tell me, and I'll find people who are."[23] Dan Steward, the team's geological coordinator, joined the company three days after Mitchell Energy drilled the C. W. Slay, and he quickly learned that George didn't like to hear that something couldn't be done. Steward sensed that George wasn't bluffing. He thought of his wife and four kids and assured George that the team would find a way to replace the depleting gas fields of the Boonsville Bend.

As the first well drilled into the Barnett, the C. W. Slay confirmed that the shale could produce gas, although many in the industry still doubted it. "I had people telling me I was a fool if I thought gas was coming from the Barnett," Steward recalled. "That was very hard for people in the industry to accept—that was hard for some people in the company to accept." George, however, believed the team was on the right track. He became increasingly convinced that the Barnett Shale could be the company's salvation.

One of the first things Steward's team did was review the data from another project Mitchell Energy had been involved in a few years earlier that used a technique known as hydraulic fracturing. Fracturing—or fracking as it would later become commonly known—dates to the 1860s, when liquid nitroglycerin was shot into the hard rock of wells in Pennsylvania, West Virginia, New York, and Kentucky. It was, of course, extremely dangerous, but it succeeded in breaking up the rock and releasing oil. In the 1930s, companies tried using nonexplosive liquids like acid under high pressure to break rocks apart. Over the years, napalm and dynamite were tried, too. The Atomic Energy Commission even tried a nuclear bomb. Working with the El Paso Natural Gas Company, it detonated at twenty-nine-kiloton nuclear warhead, dubbed the "Gasbuggy," four thousand feet underground near Farmington, New Mexico, in late 1967. The explosion left a crater 80 feet wide and 335 feet deep, and the wells drilled into it produced gas from the sandstone underneath, but it was radioactive. The commission tried two more times in Colorado, with similar results, before the idea of "nuclear fracking" was abandoned.[24]

Today, George is often erroneously called the "father of fracking." If fracking has a father, it was Floyd Farris. Farris worked for Stanolind Oil and Gas Corp., a division of the company that became Amoco. (Amoco was later bought by BP.) In 1947, Farris used napalm and sand to fracture a well in southwestern Kansas. The process didn't increase production appreciably, but Stanolind patented it anyway and later licensed the technology to Halliburton. Halliburton, one of the biggest oil-field service companies, completed its first two commercial "fracks" in 1949, one in Archer County, Texas, and the other in Stephens County, Oklahoma, using gasoline instead of napalm. By the mid-1950s, more than three thousand wells had been fracked in some way, and the fluids injected evolved from gasoline to various water-based gels made from different thickening agents, such as guar gum, the ingredient most commonly used to make ice cream creamy.[25] Mitchell Energy had used fracking over the years to stimulate wells in the Boonsville Bend, but no one had tried fracking shales.[26]

As the Mitchell team studied and tested the Barnett, an idea the rest of the industry thought was ludicrous—that gas could be extracted from shales without natural fractures—began to seem feasible. In 1979, Mitchell Energy, working with the government-funded Gas Research Institute, had pumped the largest frack job ever done at the time to boost the production of a well in the North Personville field in Limestone County, Texas. The frack, which used gelled water, doubled the well's production and helped make the field Mitchell Energy's second-biggest gas producer behind the North Texas play. The government covered the cost of the expensive gels, but even so, the well was only marginally profitable.

The project's success raised the possibility that larger frack jobs might also work on the Barnett. George told his operations staff to frack the C. W. Slay well and see whether the results improved. They didn't, but he remained optimistic. Over the next few years, they deepened several dozen more wells into the Barnett formation and fracked them. Because Mitchell Energy was still getting the higher price for its gas under the NGPL contract, the economics worked—barely.[27]

For the next few years, George's convictions would be tested as the oil and gas boom of the early 1980s gave way to a devastating bust. Companies across the industry were hit hard, Mitchell Energy included. While oil prices tumbled, the signs from the Barnett project weren't particularly encouraging. The shale was producing gas, but by the best estimates, the wells were barely breaking even.[28] Without the above-market price guaranteed under the NGPL contract, fracking the Barnett didn't make economic sense.

The company got a little help from the federal government, too. In 1978, President Jimmy Carter had signed the Natural Gas Policy Act. George had supported Carter, but he disagreed with him on energy policy. He was furious with Carter and his energy secretary, James Schlesinger, for slapping new regulations on natural gas.[29] (Though he agreed with Carter's conservation policies, he also opposed the president's windfall profit tax on oil companies, arguing that producers should be given more incentives to pour profits into boosting domestic production.) At the time, the prevailing wisdom was that the country was running out of natural gas, and prices would rise. The law put price ceilings on older gas wells like those in the Boonsville Bend, but it allowed higher prices for "unconventional" gas projects such as the frack jobs Mitchell Energy was doing in the Barnett Shale. Still, George needed more than a regulatory loophole if he was going to save the company. A decade after Mitchell Energy engineers tapped the Barnett Shale, the project was still little more than what Dan Steward would call a giant science fair project. It was far from the new lease on life that the company needed to survive.

CHAPTER 19

Fire from the Ground

In the mid-1980s, Pat Bartlett and her husband, James, scraped together enough money to buy three acres of land in Wise County, Texas, near the town of Boyd, about thirty miles northwest of Fort Worth. With the $5,000 or so they had left, they bought lumber for a home that James, a roofer, planned to build for Pat, who was wheelchair bound from an automobile accident. Like many landowners in Texas who live outside municipal water districts, the Bartletts drilled a well to supply their new home. That's when their nightmare began. As the drill bit hit the underground water reservoir, a pillar of flames shot out of the well, reaching as high as six feet in the air. It burned for six weeks and left the surrounding field scorched as if a rocket had been launched on the site.

The Texas Railroad Commission determined that the water well was contaminated with natural gas, and because of the fire hazard, it prohibited the Bartletts from using the well or drilling another. Unable to afford to move, they lived without running water for the next decade, using a five-gallon paint bucket for a toilet and hauling drinking water from town each day. To supplement their water supply, they collected rainwater from the roof, and they bathed infrequently. Classmates teased their ten-year-old son for the way he smelled. He ran away from home and later dropped out of school. Pat lived in constant fear that the well would explode again. The Bartletts' hopes of a dream house in the country had devolved into daily anguish and inconvenience, and Pat blamed one man for their troubles: George Mitchell.[1]

The Boonsville Bend may have been an ATM for Mitchell Energy, but Bartlett saw it as a curse that had tainted the water under her property. She wasn't alone. Her neighbors complained that water from their wells smelled like rotten eggs and said it left behind a black substance that corroded pipes, ruined electric appliances, and stained toilets and showers. The sour taste made the water undrinkable, and several homeowners even claimed their showers produced unsafe levels of hydrogen sulfide. They blamed the contamination for a variety of ailments, including dizziness, headaches, diarrhea, and lethargy.[2]

The Bartletts sued Mitchell Energy in 1987, the first of eight landowners in the area who accused the company of ruining the local water supply with its drilling practices. Their claims were strikingly similar to those filed decades later by homeowners around the country who claimed fracking polluted their water.[3] The Wise County suit, though, didn't involve fracking; it involved the wells Mitchell Energy had been drilling since the 1950s.

The homeowners' complaints underscored years of friction between George's exploits in North Texas and the people who worked for him and owned the land he leased. In the early days, Johnny could throw a barbecue and sweet-talk the crowd when they got restless, but those days were gone. Local landowners, eager for royalty payments, didn't appreciate George's hard-nosed negotiating skills. Over the years, he'd locked up thousands of acres, many of which he didn't drill for decades. And when he did drill, he was slow to pay royalties—a common practice among independent producers, but not one that endeared Mitchell Energy to the locals. Homer Hershey, who oversaw Mitchell Energy's North Texas properties in the late 1970s and early 1980s, warned George that by driving such hard bargains he was hurting the company's chances of expanding, because Mitchell Energy had a bad reputation in the area. Hershey, like Don Covey, had come from Shell, and both men urged George to adopt a big-company mentality and pay royalty owners more quickly. They also encouraged company representatives in Wise County to become more involved in the community. While things improved during the 1980s, the changes weren't enough to assuage landowners by the time the Bartlett suit went to trial.[4]

What's more, many Wise County residents viewed George as an absentee owner who was using their land to make a lot of money without regard for them or their neighbors. In a state as big as Texas, regional divisions are common. Mitchell Energy's headquarters in The Woodlands were about as far from Wise County as Richmond, Virginia, is from New York City, and the cultural differences were almost as great. In the county, George was viewed as a metropolitan tycoon from the state's biggest city who was draining their resources and polluting their water while filling up his own coffers down in Houston. "A lot of people here feel their quality of life is dependent on George Mitchell, and he has been jerking them one way and then back the other," the editor of the local newspaper told the *Houston Chronicle.*

It didn't help that Mitchell Energy had cycled through several local managers, each of whom had stayed only a year or two. To make matters worse, the company moved its North Texas headquarters to Fort Worth, insulting local business leaders who'd supported George's efforts in the county since the 1950s.[5] Inside the company, Wise County had become known as "Mitchell's Siberia."

Investigators with the Railroad Commission had been poking around Mitchell Energy's operations in Wise County since 1977, when a landowner complained that one of the company's gas wells was responsible for a fire that had ignited his water well. The regulators concluded that the cause of the fire, as well as several other well contamination issues in the area, was inadequate well casing. Wells are not a single piece of pipe entering the ground. They actually have pipes within pipes, and the outer pipe, known as a surface casing, is the shortest one. Its primary function is to protect freshwater aquifers from contamination by oil and gas coming up the well bore. How deep that casing needs to go depends on the geology of a specific area. After the complaints in Wise County, the Railroad Commission decided Mitchell Energy's surface casings weren't deep enough and told the company to add cement to the production string, the main conduit through which gas flowed to the surface, to protect the surface casings from corrosion that could lead to contamination of the water table. The company refused. The regulations for surface casings in the area required a depth of three hundred feet, and Mitchell Energy complied with that. Recementing even a small area of its massive Boonsville Bend field would cost more than half a million dollars, company officials argued. "If the Commission's proposal was later expanded to include the entire Boonsville field, Mitchell [Energy] would be obligated to a loss of several million dollars," the company said in a written response to regulators. Essentially, the commission wanted to punish the company for following the rules, Mitchell Energy argued. Besides, the company insisted the gas in the water was coming from natural sources. If gas were leaking from the wells, the company would see a buildup of surface pressure, and it wasn't. In the face of Mitchell Energy's push-back, the Railroad Commission backed off its demands.[6]

Back in The Woodlands, George wasn't worried about the Bartletts' claims. He felt that, scientifically speaking, they were ridiculous. Anyone who'd lived in Wise County knew that the water around Boyd had always smelled like rotten eggs. Back in the 1940s—before any gas wells were drilled in the area—visiting basketball teams brought their own water when they played at Boyd High School.[7] Residents used flares—the same technique oil companies used to burn off unwanted natural gas—to eliminate the methane from water wells fifty years before George got the call from the Chicago bookie about the Boonsville Bend.

George's attorneys were confident the company would prevail. Several similar suits had been filed before and never went anywhere because plaintiffs' attorneys had been unwilling to take a contingency payment to challenge Mitchell Energy. If these contamination claims were so clear-cut, why hadn't more lawyers been eager to take them? The Bartletts' attorneys would spend $2 million to $3 million preparing for the case.

The process for drilling a water well wasn't much different from drilling for gas, and although water wells were shallower, methane sometimes seeped into the water table naturally. Mitchell Energy drilled its natural gas wells to a depth far below the water table.

The homeowners picked up on the commission's earlier findings and claimed the cement in Mitchell Energy's surface casings did not go deep enough to protect the water table. They also claimed that independent studies found that the same elements present in Mitchell Energy's gas wells were also present in their water. The Railroad Commission inspected 259 wells in the county and found that 44 of them had various problems involving the casings or improper seals. While the discovery didn't prove that Mitchell Energy's wells tainted the water—other operators in the area also had insufficient casing—it was the sort of finding that could sway a jury.

Still, Mitchell Energy's lawyers entered the courtroom in Decatur, the Wise County seat, confident that when jurors saw the science involved, they would side with the company. The attorneys argued that the company's drilling wasn't the source of the contamination and that the county's gas-bearing sands were so shallow that almost any water well could ignite. The geology revealed zones of naturally occurring methane in the rock that seeped into the water table, regardless of any drilling activity. Two independent geologists backed up Mitchell Energy's case, and the lawyers noted that although the company had plugged all its old wells to meet the environmental requirements at the time, it had voluntarily examined them and applied new cement to the casings if needed, which it wasn't required to do.

The Bartletts' attorneys countered with dramatic photographs of burning water wells. They also hammered on the lax reporting to the Railroad Commission, saying Mitchell Energy had flouted regulations for forty years, knowing about the contamination the whole time. (Apparently, they never learned of the missing completion report from the first Lafitte's Gold well in the Galveston rail yard.) The lawyers presented internal memos that appeared to directly contradict official documents filed with the commission. But the most damning testimony was George himself. For years, he condemned oil and gas companies for not being smart about public relations, and he insisted that every oil company president should also be the head of PR. "The major companies are very smart, they're secretive, they're resource-rich, they've got tremendous organization—and they're hated," he said. "They've got a terrible image." George had always cultivated a different approach for Mitchell Energy, one in which he appeared congenial and the company appeared environmentally sensitive.[8] He'd shown, for example, that the company could drill for oil while still protecting habitat for the endangered whooping crane in South Texas. Mitchell Energy had built a new marsh using dredge material from a channel in Mesquite Bay for a new fifteen-acre habitat in the

Aransas National Wildlife Refuge, where whoopers spent the winter. The dredging gave the company access to a well site, but it also created a "Whooper Hotel" that won accolades from biologists with the US Fish and Wildlife Service for enhancing the habitat for the last wild breeding stock of the endangered birds.[9]

But now, with his own company under attack, George appeared to be just another oil executive under the gun. He may have been frustrated by the suit or impatient with the tediousness of the legal process, but in his videotaped deposition he appeared both uncharacteristically haughty and unwilling to accept any responsibility. The company's actions complied with all regulations, he insisted, and by the way, didn't anyone realize that he had rescued the entire county from economic despair? Normally relaxed and open in front of the camera, George appeared fidgety. He grimaced and acknowledged that he hadn't focused on the contamination issue or how his company plugged wells in Wise County until the lawsuit forced him to. The lawyers, he argued, should "go talk to the people involved." Other than reading the production reports, George never concerned himself with such details, but his response seemed like a brush-off by an executive too busy to care about the harm his company caused. He denied any knowledge of the internal memos or the conflicting Railroad Commission filings. It got worse. At one point, his defensiveness bordered on petulance. "Why are you hammering on me?" he demanded.

The videotape had been edited by the plaintiffs' attorneys to make George look bad, and the company attorneys later got the unedited version shown in court, but it was too late. George had come off as just another greedy oil executive, exploiting poor landowners and shipping his profits down to Houston so he could build a fancy new suburb for rich people. The apparent combination of corporate arrogance and irresponsibility proved devastating. Many of the jurors were landowners themselves, and they not only wanted the pollution stopped, they wanted to send Mitchell Energy a strong message.

After hearing three weeks of testimony, the jury in late February 1996 unanimously sided with homeowners. It found that Mitchell Energy had acted with gross negligence and malice in contaminating the county's water. Jurors awarded homeowners $4 million in actual damages and a whopping $200 million in punitive damages. The size of the judgment—more than five times Mitchell Energy's earnings the previous year—was stunning given the nature of the dispute. While the landowners had complained about health problems, those claims weren't asserted in the lawsuit. One analyst pointed out that there were no sick people "or even sick cows." The punitive damages were based in part on the jury's finding that Mitchell Energy had lied to the Railroad Commission about leaks from the wells.[10] One juror later said of the company: "They were here to make all the money they could."[11]

George was furious, and his employees in Wise County were shocked. Jay Ewing hadn't been following the court proceedings too closely. As a Mitchell Energy field manager in Wise County, he didn't give a lot of credence to the landowners' claims. Late on the night of the verdict, he got a call from his supervisor telling him they had lost the case. When he heard the amount of the damages, Ewing assumed he needed to look for another job. The company, he figured, was finished. "I felt sick," he said. "We're done."[12]

George wasn't ready to give up. The company fired out a statement in his name: "I have never believed, nor do I believe now, that Mitchell Energy Corporation is the cause of the problems that the plaintiffs are complaining about." That, at least, was the official view. But George was angry at the outcome, and he saw the people of Wise County as ungrateful for the economic benefits he had brought to the area. His anger came through in an interview in which he told a reporter that Wise County had been a "burned up, parched miserable place" before his success with the Frustration Fields four decades earlier had brought an economic windfall to local landowners.[13] The comment enraged residents even more, but his sentiment wasn't unusual in the oil business. Many in the industry believe the public doesn't appreciate how the industry consistently delivers affordable energy and pays royalties to local landowners. Still, George was typically more diplomatic. In this case, his willingness to speak openly with the press, and the lack of media gatekeepers at the company worked against him. But he knew that such a large verdict could wipe out the company's stock value. "He was troubled by the thought that he and the company would be devastated by it and his plans would be destroyed," Dancie Ware said.

◈

In 1991, Pat Bartlett was sitting in a Fort Worth café when an underground gas explosion blasted customers out of their chairs. She ultimately received a $100,000 settlement, which, by the time of the Wise County verdict, she and her husband had used to buy a new house that was connected to a municipal water system. They stood to receive $13 million from their share of the Mitchell Energy judgment.[14] Ironically, even though the plaintiffs had accused Mitchell Energy of contaminating the groundwater, none of the money awarded was designated for environmental cleanup. The attorneys representing the Bartletts would get $80 million for taking the case on contingency. And there was another twist. The judge in the case, John Fostel, had previously worked for the plaintiffs' law firm. He had an agreement that allowed him to continue to collect 20 percent of any contingency fees from cases the firm was working on when he left, which applied to the Bartlett matter. Mitchell Energy's attorneys also pointed out that the judge had owned an eighty-acre tract in Wise County that drew water from the same aquifer the Bartletts had accused Mitchell Energy of polluting. Although he had turned over the land to his ex-wife by the time of the trial, Fostel and his family could potentially have filed their own claim against Mitchell. The *Wall Street Journal* lashed out at the whole arrangement in an editorial in June 1996 in which the newspaper held the case up as an example of runaway Texas juries and the need for tort reform.[15] (Texas had enacted sweeping tort reform legislation in 1995, but it didn't apply to the Bartlett case, which had been filed before the changes took effect.)[16]

Mitchell Energy's lawyers immediately asked for a new trial. But the judge—Fostel had recused himself by then—denied the request. The only option was to appeal to the state appeals court in Fort Worth. At the appeals level, there was no jury to get caught up in the emotions of homeowners using buckets for toilets. It would be a rational discussion of the facts, and George wanted to make sure there was no room for doubt. Instead of simply presenting geological theories, they tested the water wells used by the Bartletts and others.

The stakes were huge. Thirty-one other homeowners had filed similar suits that were queued up in state court waiting for trial dates. Not only did the homeowners claim the water tasted bad, they blamed the contamination for an even broader variety of health issues—fatigue, depression, dizziness, nausea, headaches, memory loss, and gall bladder problems.

Mitchell Energy faced a legal deluge, any part of which would have pushed the company to the brink of bankruptcy. As news of the jury's decision hit Wall Street, the stock price began to fall from more than $18 to as low as $16.38. The question of how Mitchell Energy would survive could no longer be relegated to the future.

Analysts renewed their demands that George split the company. As COO, Bill Stevens became more convinced than ever that they were right. George knew something had to be done, but now his hands were tied. He had loved the energy business since he had first worked for Johnny in southern Louisiana before college, but it had always been a means of making money. The Woodlands, on the other hand, was an entire community he had created from his own imagination. It was, as he liked to say, his eleventh child. Yet faced with the jury verdict and the incessant financial pressure on the company, he didn't have any choice. He could no longer ignore the reality that the company needed to be either a real estate firm or an energy concern. And with the verdict hanging over it, selling the energy company was impossible. "No one wants to buy a lawsuit," George said. But the obvious logic of what needed to happen didn't make the decision any easier. The Woodlands had to be sold.

◈

Within days of the Bartlett verdict, George and Bill Stevens met with Houston attorney Jack O'Neill and hired him to take over the case. Mitchell Energy needed a new legal strategy. The verdict in the Bartlett case touched off a wave of new lawsuits and a new round of regulatory scrutiny. O'Neill called the jury's decision in the Bartlett case "preposterous" and predicted it would be overturned on appeal.[17]

Meanwhile, a second case involving seventeen more homeowners was set for trial in a few months, and more suits were pending.[18] The company faced countless years of litigation as it staggered under the Bartlett verdict. By early 1997, Mitchell Energy was a defendant in twenty-nine other cases in Wise County over alleged water contamination.

The next lawsuit, known as the Bailey case for one of the homeowners who filed it, went to trial in March 1997.[19] This time, neither the company nor its legal team was taking anything for granted. George and Bill Stevens were determined that Mitchell Energy wasn't going to face a repeat of the Bartlett debacle. O'Neill had a three-pronged strategy. He wanted to know whether residents of Wise County could ignite their water before any drilling had begun in the region; he wanted to know whether gas could be chemically typed to determine where it came from; and he suggested that Mitchell Energy offer to provide the plaintiffs with a system that would remove all gas from their water supply.

Homer Hershey and other Mitchell Energy employees who were going to testify at the trial went through days of training in Fort Worth, learning how to conduct themselves on the witness stand and avoid showing emotions that could turn the jury against them.

During the trial, O'Neill presented testimony from longtime Wise County residents who said the water had been flammable years before Mitchell Energy tapped the Boonsville Bend. He also produced an expert who showed that gas could indeed be chemically typed, and that the gas present in the Wise County water wasn't coming from Mitchell Energy's wells. Many water wells were in fields used by livestock and weren't drilled or sealed properly to keep out runoff. The tests showed that the methane causing the water to smell was biological. It wasn't coming from thousands of feet underground and leaking upward; it was coming from the surface—from cattle feces—and soaking downward. And finally, O'Neill noted that none of the plaintiffs accepted Mitchell Energy's offer of a water purification system, which, in the minds of the jury, made it appear they were more concerned with money than with solving the contamination problem.[20]

Testimony in the trial lasted eleven and a half weeks. Every day after court, the legal team, Hershey, and others from the company who were involved in the case met at a Residence Inn on the outskirts of Fort Worth. They discussed what had happened in court and planned for the next day.[21] The jury deliberated for less than five hours before ruling in Mitchell Energy's favor. The decision was a stunning reversal from what had happened the previous year in the Bartlett case, and it offered hope that Mitchell Energy might yet survive the litigation.[22] By finding that the company didn't contaminate the water, the decision "should put to rest many of the false allegations arising from an earlier verdict in a similar case," George said in a company-issued statement. The decision, Mitchell Energy claimed at the time, indicated it would be vindicated in the Bartlett case as well.[23] Six months later, it was. The appeals court reversed the Bartlett verdict, removing the threat of the $204 million verdict.[24] The company still faced more than two dozen other lawsuits making similar claims, but with two solid court rulings now in Mitchell Energy's favor, the plaintiffs in those cases were willing to cut a deal.

As the legal proceedings ground on in court, Railroad Commission investigators once again scrutinized Mitchell Energy's well casings in the area. They found that 112 wells in Wise and three surrounding counties had casings too shallow to protect groundwater. In a 1998 report, the commission claimed the company had "deliberately misreported" the depth of protective well casings for 110 of the wells. Eight others hadn't been plugged properly when the company stopped producing gas from them. The commission ordered Mitchell Energy to pay $2.24 million, the biggest fine in the agency's history.[25]

By September 1998, Mitchell Energy had settled all the remaining lawsuits. Although the terms weren't revealed at the time, the company agreed to pay the landowners $21 million in higher royalties through 2008.[26] Two months later, it offered to settle the Railroad Commission's fine for $100,000. It cited the jury verdict in the second case that found the company wasn't responsible for water well contamination, and it noted, once again, that it had done more than was required to address the problems. It had spent $2.8 million to repair bad well casings that it had identified, another $3.6 million for a water pipeline from Lake Bridgeport so residents could hook up to municipal water rather than rely on wells, and another $450,000 on treatment systems for residents who remained on well water. Once again, the commission relented. The company paid $200,000 to settle the investigation.[27]

The litigation had been costly, but in forcing the decision to split the company, it did some good for Mitchell Energy's long-suffering stock price. By the end of 1997, the shares were trading at more than twenty-nine dollars, or 45 percent higher than after the Bartlett verdict in February 1996. The Woodlands wasn't the only casualty. The company also sold off some of the forestland around Cook's Branch. The San Luis hotel in Galveston, which had opened just a decade earlier to such fanfare, was sold to Tilman Fertitta, an island native, owner of the Landry's seafood restaurant chain and great-grandnephew of Sam and Rose Maceo. (Years later, after building Landry's into a national empire, Fertitta would gain fame hosting the cable television reality show *Billion Dollar Buyer*.)

Mitchell Energy was now what Wall Street had always wanted it to be: a pure energy play. George had paid a high price, selling his beloved Woodlands. If he had been superstitious, he might have noted that news of the sale, in June 1997, hit the front page of the *Houston Chronicle* on Friday the thirteenth.[28] While it pained George to sell The Woodlands, many in the energy side of the business were relieved. In hindsight, Dan Steward said the lawsuits were "a godsend" because they prevented George from selling the energy business before he had proven what fracking could do in the Barnett. As difficult and devastating as the legal battle had been, the biggest challenges for George and the company still lay ahead.

CHAPTER 20

The Frack Heard Round the World

While the attorneys battled the lawsuits in North Texas, another pressing company matter captured everyone's attention. For four decades, the wells in Wise County had been a cash cow for Mitchell Energy. By the mid-1990s, the firm's annual revenue topped $1 billion, with one-quarter of that coming from the North Texas leases. For most of that time, few other companies could copy Mitchell Energy's success, because it had something they didn't: the NGPL contract. Thanks to George's negotiating skills, NGPL, which by then was owned by Occidental Petroleum, still paid higher prices than natural gas fetched on the spot market to guarantee the supply for its customers. Now, it was becoming clear that predictions of surging gas prices were woefully off base.

In Dallas, the legendary oilman and one-time corporate raider Boone Pickens had bet big on natural gas and lost. In the mid-1980s, he took Mesa Petroleum, the company he had founded in 1958, and used it as a takeover vehicle for companies such as Gulf Oil, Phillips Petroleum, and Unocal. Then, he turned Mesa into a partnership focused on natural gas. Unlike companies, partnerships distribute most of their cash to investors. Pickens was offering investors a chance to participate in the rise in natural gas prices that he—and many other analysts and forecasters—saw as inevitable. Instead, gas prices fell, and Pickens had to borrow $1 billion to pay shareholders. Mesa never recovered. In 1996, the company's deteriorating finances forced Pickens to sell it, and the new owners ousted him. "Natural gas—hell, it cost me my job at Mesa," recalled Pickens, who went on to make an even bigger fortune in the early 2000s as an energy trader and a hedge fund operator.

George had insulated his company from the vagaries of the

gas market largely because he knew he could sell his production to NGPL at a high price. The liquids plant enabled him to strip other elements from the gas and sell them separately. But he also knew that unless the company could lower its Barnett well costs or increase its reserves in Wise County, he might eventually suffer the same fate as Pickens.

By the time of The Woodlands sale in 1997, Mitchell Energy had been drilling into the Barnett Shale for more than fifteen years. It had created and settled the debate over whether gas could be extracted by fracking—it could. But it was too expensive. Because of the special chemical gels used in cracking open the shale, the fracking process often cost more than the drilling.

Now, the company had a new problem. In mid-1995, NGPL sent Mitchell Energy a letter saying it intended to terminate the contract it had first signed with Oil Drilling Company in the early 1950s. It would make a series of payments totaling almost $242 million to end the agreement as of January 1, 1998.[1] Like so many other natural gas producers, George now had a gun to his head. But unlike so many of his rivals, George believed the solution was within the company's grasp. All it had to do was cut the cost of developing the Barnett more than anybody deemed possible. "Everybody thought he was out of his mind with what he was doing in North Texas," Pickens said. That included Bill Stevens, who saw the company lurching closer to crisis, while George continued to entertain dreams of a solution that simply didn't make economic sense.

"Everybody knew you did not produce gas from a source rock," said Homer Hershey, the longtime Mitchell Energy engineer who oversaw the Barnett Shale development. Hershey was among the doubters inside the company initially, but over time, he changed his mind as he saw the improving results from test wells. While fracking the Barnett was expensive, the production from the five thousand shallower wells that the company had drilled in the Boonsville Bend covered the costs of the tests. Rather than drill exploratory wells, Mitchell Energy started its experiment by extending the existing Boonsville wells deeper into the shale, which kept expenses low. These wells were commercially viable because of their shallower reserves, and they were already connected to a gas-gathering system, so the flow of gas could be monitored without additional equipment.

"We pretty well had a laboratory in the field to do some testing," Hershey said. "That was the key to the company's success."

In drilling new wells into the Barnett, the engineering team worked close to the natural faults that already existed in the shale. In theory, that would require less fracking because the shale was already split. "We couldn't have been more wrong," Hershey said. The gas near the faults was depleted because it flowed from there up to the Boonsville, where Mitchell Energy had been pumping it out for decades.[2]

Unlocking the secret of the shale would require a different approach.

◈

The breakthrough that George sought for so long began at a baseball game. In the fall of 1996, the Texas Rangers were playing at the major-league park in Arlington, a suburb between Dallas and Fort Worth. Nick Steinsberger had gone to the game with some fellow petroleum engineers from different companies operating in the Fort Worth Basin.

Steinsberger was in his early thirties and overseeing the completion of the Barnett Shale wells for Mitchell Energy. The company had been through a couple of rounds of layoffs in the 1990s, and he was increasingly worried that he might lose his job.[3] After all, many Mitchell Energy insiders thought the Barnett project was a waste of time. Then, things got worse. By the end of 1998, crude oil was selling for less than $11 a barrel, having fallen 58 percent in two years.[4] Natural gas experienced a similar decline, losing half its value during the same period and hitting a low of $1.72. Even the biggest companies were struggling. Within a few years, major international oil companies like Exxon and Mobil, BP and Amoco, and Conoco and Phillips would merge. Smaller ones—many, in fact, the size of Mitchell Energy—would be swallowed up or liquidated.

While these concerns weighed on Steinsberger, they really weren't anything new. He had spent his whole career in one of the modern energy industry's bleakest chapters—the late 1980s to the late 1990s. He graduated from the University of Texas at Austin with a petroleum engineering degree in 1987, right in the teeth of a major oil bust. Companies simply weren't hiring engineers in the United States. Steinsberger's college roommate was from Norway and took a job with that country's state-owned oil company, Statoil, in London. Steinsberger, with nothing else to do, decided to tag along. He spent about ten months traveling in Europe before coming back to Austin to pick through prospects at the university's placement center. This time, he found one company that was hiring—Mitchell Energy and Development.

The job wasn't glamorous. He would be working on two small well sites near Whittier, California, a suburb southwest of Los Angeles. "I was engineer, pumper, director, foreman," he said. Less than a year later, the company transferred him to the Bridgeport office, and after a year and a half in a training program that gave him a taste of the different aspects of natural gas production—drilling, completions, facilities—he began working in 1989 as one of ten field engineers in "Mitchell's Siberia."

Most production in the Fort Worth Basin still came from the shallow wells in the limestone of the Boonsville Bend. Only between fifty and seventy had been drilled into the Barnett Shale and fracked. "We were poking keyholes in the Barnett," Steinsberger said. Initially, he oversaw production from about five hundred of the shallower wells, doing what he could to increase or at least maintain production, so the company could meet the terms of the NGPL contract. The routine continued for about five years.

The Barnett wells had their own engineer to monitor them, but it wasn't a prestigious position. The wells barely produced enough to justify their cost—and then, only because the company needed gas to supply its processing plant in Bridgeport. As a result, engineers on the project changed every few years. In 1995, the latest Barnett engineer got promoted to the home office in The Woodlands, and Steinsberger, who was one of the youngest engineers in Bridgeport, was given the job.

He drove to his first meeting with managers in the reservoir engineering group and was surprised by their assessment of the Barnett project. "I was told that we would probably drill another 30 or 40 wells and give up because the Barnett wasn't working. It was not economic," Steinsberger said. "I had just gotten this promotion. It was a bit of a let-down." The senior engineers believed the company needed another solution to address the depletion of the Boonsville Bend. (The geology team, however, didn't agree with their assessment.)[5]

Steinsberger returned to Bridgeport and decided that all he could do was reduce drilling costs as much as possible. But the special gels used in fracking pushed the expense of a typical well to $750,000 or $850,000. About 40 percent of that was the frack job itself, and the rest went for a couple million pounds of sand and the high-viscosity gelled fluids that carried it into the fractures and held it in place. Steinsberger focused his savings program on the priciest of the ingredients—the gels.

When a well is fracked, the gels carry the sand out into the fractures and then break down over time, leaving the grains of sand to keep the fractures open so the gas can flow through. Conventional wisdom said that the gel ensured the sand was carried all the way into the fractures. With less viscous fluids, the sand would simply settle in the bottom of the drill pipe, a problem known as "screening out."[6]

The oil-field service companies that Mitchell Energy hired to provide the gels sold them by the gallon—at markups as high as 1,000 percent—so Steinsberger decided to cut back on the amount of gel per well and see what happened. The service companies warned him that the high temperatures of the Barnett would cause the gels to break down and they wouldn't distribute the sand to the fractures, but he didn't notice any change in production from the cutbacks.

While the change didn't save enough money to make the Barnett wells profitable, Steinsberger began to question the thinking about gels. That's when he wound up at the Rangers game in the fall of 1996 with a couple of engineers from Union Pacific Resources, another Fort Worth–based oil company that had been spun off from the railroad of the same name years earlier. Unlike in major energy hubs such as Houston or Midland, the petroleum engineering community in Fort Worth is small. Most of the engineers know each other, and it isn't uncommon for them to share ideas or problems with colleagues at other companies.

One of the UPR engineers mentioned that his company was doing something called a "slick-water frack" in an East Texas formation known as the Cotton Valley. What they were doing wasn't supposed to work, but the UPR engineers claimed to have cut the cost of fracking in half while getting good production from the wells. No one knew why it worked, but it did—at least in limestone. Limestone was far more porous than shale. Could the technique also work in the Barnett?

Steinsberger was skeptical, but he met with the engineers at their office a few days later to learn more about water fracks. Given the high cost of fracking the Barnett and the need to find new reserves to keep the company in business, he decided he didn't have much to lose.

At a meeting in The Woodlands in late 1996, he made his case to Mitchell Energy executives. He proposed a series of wells using the new slick-water technique. If it didn't work, he argued, the company could always refrack the wells with gels. All they would lose was a little time and some water. The reaction in the room was mixed. "Some people thought I was full of shit to even try something like that and other people were on board," Steinsberger said. "It wasn't like 'this is an a-ha moment and it's a great thing to try.' It was a risk."

However, Dan Steward, who was coordinating the Barnett project, supported the effort, as did Mark Whitley, who was Steinsberger's immediate supervisor.[7] The management team allowed Steinsberger to drill three wells using slick-water fracks in the spring of 1997. Any optimism, however, was short-lived. Each of the three wells screened out. The water didn't carry the sand far enough from the well bore, and pressure spiked. Steinsberger stopped the operation early to keep the production casing in the well from bursting.

He managed to complete the wells, but after monitoring their production for several months, it was about the same as if he'd used gel fractures. "They were all failures," he said. After poring over the results, he realized his mistake. He'd basically taken UPR's recipe for the East Texas formation and applied it to the Barnett. "That was stupid in hindsight," Steinsberger said. The East Texas limestone that UPR fracked was more permeable than shale. "Pumping slick-water fracks with even a little bit of sand is hard to do in shale," he said. "I didn't know that at the time."

The realization sent him back to The Woodlands in the winter of 1997 to plead for another chance. Senior management agreed to give him three more wells to continue the experiment. For one of them, the S. H. Griffin #4, Steinsberger tried a different technique. Rather than adding all the sand at once, he started with a smaller amount and gradually increased the concentration, hoping that it would prevent the screening-out problem. He continued to monitor the results and compared them with those of other wells in the area.

Each well had a different sand concentration. He kept charts that rated each well "A" through "F," based on how he fracked it, how it produced, and other criteria. An "A" well might produce a million cubic feet of gas a day for the first ninety days. When it was completed, the Griffin well was off the charts. It wound up averaging a million cubic feet a day for more than a year. "That was a huge success," he said. "The S. H. Griffin wound up being the best well that we'd ever completed at that time."

If there was an "a-ha" moment in fracking's evolution, this was it. Steinsberger had done the impossible—he'd figured out how to frack shale commercially. His technique for gradually adding sand to the water is still used in most fracking jobs today, but at the time, he had no idea of the magnitude of his discovery. All he knew was that he had found a solution to the cost problem that had plagued Mitchell Energy. By using water instead of gels, he had cut total well costs by about 30 percent on each well.

During the next two months, all three wells showed signs of success. In another six months, by late 1998, Mitchell Energy was using slick-water fracks exclusively to complete the Barnett.

Throughout the entire project, Steinsberger estimates he probably had no more than ten meetings with George, who monitored much of the progress through Steward and Whitley. After the success with the Barnett, George didn't call the one person most responsible for the breakthrough that saved his company and, ultimately, changed the world of energy.

It wasn't a snub so much as a reflection of George's discomfort with intimate conversations, especially with people he didn't know well. "If you knew him, you just knew that he was not the most verbally effusive guy," his son Todd said. "Yet I know my father had the greatest regard for the people who did the detail work. Sometimes simple messages that people would expect to communicate with each other would get lost in translation."

It may also have been that the implications of Steinsberger's breakthrough took a while to manifest. Even a year after the S. H. Griffin came in, Mitchell Energy was still laying off employees. "It wasn't enough to turn the company around yet," Steinsberger said.

◈

In late February 1998, Kent Bowker walked into the break room at Mitchell Energy's offices in The Woodlands to grab a cup of coffee. Bill Stevens was already at the pot. Bowker had joined the company two days earlier and walked over to introduce himself. He told Stevens how excited he was to be working for Mitchell and how much enthusiasm he had for the Barnett Shale. He was convinced the play held far more gas than anyone, including the company's executives, realized. Stevens put his hand up. "Stop right there," he said. "We've had enough of that stuff."

Bowker was stunned, and Stevens left him standing alone in the break room wondering whether he'd made a terrible mistake in the joining the company. "I've got all this enthusiasm, and here's the president of the company sticking his hand in my face, telling me this isn't going anywhere, so why are you wasting your time on it," Bowker said. Bewildered, he walked upstairs to Dan Steward's office and asked what he'd gotten himself into. "Don't listen to him," Steward said. "George Mitchell's behind the Barnett. It's OK."

Bowker, who's stocky and has short gray hair and John Lennon glasses, had landed right in the middle of the growing rift between George and Stevens over fracking in the Barnett Shale. What Bowker had discovered would only make George more determined that the Barnett would save the company.

Bowker, whose voice tends to rise an octave when he's excited, was convinced he was right. Before joining Mitchell Energy, he'd spent years studying the Barnett while working for Chevron. The oil major had studied US shale formations in the early 1990s, and by the middle of the decade, it had bought property in Johnson County, not far from Mitchell Energy's Barnett leases. Bowker pored over the Barnett data that Mitchell Energy shared with the Department of Energy and the Gas Research Institute. The numbers looked promising, but Bowker had done a lot of research into the nature of shale formations and how they held hydrocarbons. To him, Mitchell Energy's estimates for the Barnett seemed low. The company developed its projections by comparing the Barnett to the Appalachian shale, which had been the focus of most of the GRI studies. But there were many inconsistencies between the two shale formations. This convinced Bowker that Chevron, a much larger company with far greater financial resources, would beat Mitchell to the solution for profitably extracting gas from the formation.

In the fall of 1997, Bowker took a core sample from a newly drilled Barnett well and put it in a desorption canister, a foot-long metal tube with rubber caps on either end. The test was designed to show how much gas was in a cylindrical core sample of shale drilled from deep within the Barnett formation. The technique wouldn't work in reservoirs with more porous rock because the gas would escape before the sample was brought to the surface. But in blackboard-like shale, the sample would retain much of the gas. Bowker believed the canister would give a more accurate picture of how much gas the Barnett held.

After the sample had been in the tube for a while, Bowker noticed that the rubber end caps were ballooning to the size of volleyballs. He ran the tests and measured far more gas than Mitchell Energy or GRI or anyone else thought was possible. "There was way, way more gas in these shales than we thought," he said. All Chevron had to do was figure out how to get it out.

Bowker didn't know about Nick Steinsberger's experiments with slick-water fracks, but it didn't matter. Soon after he confirmed the test results, a new management team at Chevron shut down the shale project. It may have held huge potential, but for a major oil company, the returns weren't big enough to justify the investment. Chevron looked for giant fields, known in the industry as "elephants." It increasingly found them only in foreign countries or far offshore. If Bowker wanted to advance his career within the company, he would have to go overseas, probably to West Africa. The idea didn't appeal to him, especially with a newborn son at home, so he decided to move to Mitchell Energy. He already lived in The Woodlands, and he had long admired George. Even more appealing, at Mitchell Energy, he could continue the work he had started at Chevron—or so he hoped.

"I was sitting on information that nobody else had," Bowker said. "You do the volumetrics, and there was more gas than the US will use—ever." Dan Steward knew Bowker's work and wanted his expertise. Bowker was outspoken, with a sarcastic streak that didn't sit well with some of his new colleagues, especially because his message was that the company had underestimated the size of its own reserves. However, he convinced Steward and Steward's boss, John Hibbeler, that his numbers were right. Steward and Hibbeler realized the implications—the Frustration Fields were on the verge of a second life, one that could prove far more prosperous than the first. If the Boonsville Bend had been an ATM, the Barnett Shale was a mint, a veritable gold mine of gas.

Outside of the Barnett team, the project's potential was lost on many fellow employees. For more than two decades, Mitchell Energy workers had watched The Woodlands siphon off energy profits. Now, George had a new pet project that many saw as another money pit. The company had endured a difficult few years with the lawsuits in North Texas, persistently low natural gas prices, and three rounds of layoffs. In 1995, it employed 1,300 people. Three years later, its ranks had dwindled to 800.[8]

During his time at Mitchell Energy, Bowker didn't have many interactions with George. But soon after he presented his theory to Steward and Hibbeler, he found himself laying out the details in a conference room with George, Bill Stevens, and Todd Mitchell lined up on one side of the table. As he worked through his analysis, Bowker addressed his comments to George. He already knew how Stevens felt. He methodically described the case for why he thought the Barnett held almost four times as much gas as Mitchell Energy had thought it did. He knew that George didn't always follow the technical details in presentations, but when he concluded his remarks, he saw a smile work across George's face as the implications of Bowker's presentation sank in. George looked at Bowker, still smiling. "This is huge," he said.[9]

CHAPTER 21

Persistent Doubts

After hearing Bowker's presentation, Bill Stevens didn't share George's enthusiasm. To him, the effort to unlock the Barnett still looked like a waste of the company's money. After seventeen years of fracking in North Texas, little had changed. Stevens didn't think the Barnett looked like salvation; he thought it looked like a costly indulgence the company could no longer afford. George had built Mitchell Energy on debt, but it had reached its borrowing limit.[1]

Company lore, viewed through the lens of fracking's future success, casts Stevens as the shortsighted, bean-counting executive who sought to crush George's dreams. But Stevens's opposition to investment in the Barnett took shape over time. Initially, Mitchell Energy continued to invest in engineering and development, and it even increased Barnett production under Stevens. Dan Steward, Nick Steinsberger, and the rest of the Barnett team had proved gas could be extracted from the shale, and Stevens didn't argue the point.

As the controlling stockholder, George could have simply demanded that the fracking research continue. He certainly encouraged his engineers and geologists to keep working regardless of Stevens's concerns. But by the mid-1990s, George was spending less and less time at the company. By 1998, he was almost eighty years old and had retreated from day-to-day operations. He worked mostly from his private office in the Chase Bank tower in downtown Houston rather than The Woodlands offices of Mitchell Energy. His small suite had few of the trappings of executive ego—no plaques, no awards, no pictures with presidents adorned the walls or cluttered the credenza. George sat behind a black marble desk, and the office decor was sparse and modern—gray walls and rugs, chrome and glass tables.[2]

Once a month, Bill Stevens, usually with Todd Mitchell in tow, would go downtown for lunch with George at the Houston Club. They discussed the company's activities, but George was no longer the dominant presence he had been.

Stevens, meanwhile, spent a lot of time cultivating the board. As a veteran of Exxon, one of the biggest corporate bureaucracies on earth, he knew how to work office politics in a way that George had never learned. George had run the company by telling directors what he wanted and nothing more. Stevens updated them on all aspects of the operations and warned them about potential financial pitfalls, the drag on financial performance from having separate real estate and energy divisions, and the poor economics of expanding the Barnett Shale experiment. Despite his doubts about the viability of that project, though, Stevens allowed it to continue. Perhaps he acted out of deference to George, whom he respected. Besides, Stevens was far from the only doubter in the company, and the efforts to develop the Barnett were not a covert program hidden from his view. The question became one of cost relative to results, and over time, the results simply weren't keeping up.[3]

As Kent Bowker wrapped up his presentation, Todd Mitchell, sitting at the other end of the table, was also less excited than his father. Of George's seven sons, Todd had worked most closely with him over the years and was only the one to pursue a degree in geology. After college, he went to work for Howard Kiatta, who left Mitchell Energy in the late 1970s to start his own company. Todd later struck out on his own, but he accepted his father's offer to join Mitchell Energy's board. During most of the 1990s, though, Todd found himself in an awkward position between George and Bill Stevens. Stevens seemed to view Todd as a conduit for getting through when George didn't want to listen to anyone else.[4]

With the sale of The Woodlands, George seemed to become even more intransigent on certain issues, especially the Barnett Shale. Despite Stevens's opposition, George continued to push the Barnett team. But with natural gas prices weakening, the question was no longer whether there was gas in the Barnett, or whether Mitchell could extract it; the question was whether the company could afford to continue the effort. The numbers simply didn't make sense. "Drilling a vertical well in the Barnett Shale in 1998 would almost guarantee you a marginal rate of return—below the rate at which you could grow the company," Todd said.

George didn't seem to care. Ever the dreamer, he believed they would make the economics work somehow. Stevens, though, saw the numbers moving against the company. Selling The Woodlands had removed the real estate drain on earnings. Now, George was determined to plow more money into the marginal shale experiment, emboldened by Bowker's prediction that a major find lurked in the Barnett. For years, Stevens had been warning the board about the economic consequences of losing the NGPL contract. Now that had happened, and the company had done little to prepare for it. If gas prices stayed at their current level or kept falling, Mitchell Energy was headed for disaster despite all the advances with fracking and the sale of The Woodlands.

Stevens had little luck getting through to George, so a bizarre form of shuttle diplomacy emerged. He would call Todd and say, "I need to meet with your father." Todd would arrange a get-together, sometimes at George's office in downtown Houston or sometimes at his beloved Houston Club. Stevens would arrive with PowerPoint slides and other materials, almost as if he were preparing for a pitch to Wall Street analysts. He would go over the financial implications of the Barnett program in methodical detail—focusing on the constraints of the finances and the geology. Rather than advocate a complete shutdown of the Barnett, Stevens stressed the economic trade-offs. Continuing to develop the core area of the Barnett, where the company had already drilled about eight hundred wells, made more sense than expanding the program to areas where they weren't as sure the geology would yield results.[5]

The problem was that neither Stevens nor Todd could tell whether George was really listening. George had a notoriously short attention span when it came to details, especially when those details were telling him something he didn't want to hear. Todd realized his job was to listen to what Stevens was saying and then try to reinforce the message with his father later. If he went to the family home on Galveston's Pirates Beach, for example, he might casually bring up the earlier discussion they had had with Stevens. The debates over the Barnett came down to George's faith versus economic reality. It was a little like saying that there's $50 trillion worth of gold floating in the oceans of the world. It doesn't mean it can be collected for less money than it costs to find it, but you know it's there. For George, the presence of gas in the Barnett was all he needed to hear. Although he ran a publicly traded company, he didn't care about next quarter's earnings, and especially not next year's. Wall Street had never understood his vision, and neither did Bill Stevens. Even Todd, as he played the role of subtle persuader, was mystified by his father's stubbornness. How could he be so sure the company would prevail?[6]

George and Stevens defined success in different ways. Stevens was trying to run a company in a financially responsible manner. George was fixated on his vision of a better world and what it would mean for the company if he could unlock the potential of the Barnett.

Todd had grown up with his father's obsessions—The Woodlands, the Aspen development, the Galveston restoration. He knew that the prospect of losing money, even for decades, wouldn't deter George. But as Todd looked at the company's situation, he realized the stakes were much greater than simply a few more years of losses. In 1998, the entire US natural gas industry faced declining prices and production. That was putting pressure on all of Mitchell Energy's operations, not just the 25 percent of the business that was tied to the Barnett. Companies that wanted to keep growing had to either buy rivals or invest in deeper, more complex drilling projects that posed a greater financial risk. Mitchell Energy didn't have the capital to go on a buying spree. George's love of leverage had already saddled it with a lot of debt. Although the sale of The Woodlands had netted about $481 million in proceeds and the company had repaid about $286 million in loans, it still had another $414 million in debt hanging over it. In 1998, its annual revenue fell to $791 million from $907 million a year earlier, and it reported a $35 million loss.[7]

Mitchell Energy was no longer the wildcatting outfit that Oil Drilling Company had been. It had found prolific and lucrative deposits of natural gas (as well as oil), mostly in Texas, and it had concentrated on exploiting them. But for the company to continue to attract investors to its stock, it needed to show that production and reserves were growing, and that was becoming costlier. New technology such as 3-D seismic surveys—modern, more accurate versions of the technology George had used to find the Lafitte's Gold field in Galveston—improved chances of finding oil and gas deposits, but it was expensive. The investment added millions to the company's capital outlays in 1998 and 1999.[8] While the capital demands were rising, the price at which Mitchell could sell its gas was falling. Other public companies might have been able to offer additional stock or tap the debt markets, but George still controlled almost two-thirds of the shares, and he refused to sell them. And because he held so much stock, Mitchell Energy's trading volumes were low, hurting its liquidity. The share price, which had started the year at more than twenty-eight dollars, had tumbled below ten dollars by the end of 1998.[9]

The stock's decline posed another danger. Most of George's personal net worth was tied to his shares—and he had pledged them as collateral for a host of charities he supported. He had become one of the highest-profile philanthropists in Houston, a city known for spreading its oil wealth around. He and Cynthia gave generously to a wide range of causes. George had established a venture capital fund to underwrite promising biotech research firms, and he and Cynthia funded a $7 million endowment for the Houston Symphony.[10] Like many rich executives whose net worth was linked to their companies' equity, George borrowed against the value of the shares. In turn, his bankers, realizing that the stock would fluctuate, extended him credit based on a range in the share price. If the stock price dropped below a certain level, the banks would demand George either pledge more shares as collateral or repay the loans. Known as a margin call, this situation would have set off a personal financial crisis for George by forcing him to sell stock. Because he was the biggest shareholder in Mitchell Energy, those sales would, in turn, further depress share price. If the market got even a hint that George's bankers wanted more collateral, the stock could plummet. The company could issue more shares, of course, but that would dilute the holdings of other investors and cause the stock to decline further. It would also dilute George's controlling stake, which he wouldn't allow. The company, its investors, and George were just a few dollars away from a dangerous downward spiral that, once begun, would be irreversible.

George had also borrowed heavily to fund the restoration of the Strand in Galveston and to purchase the Hotel Galvez. None of the properties made money. All told, George's personal debt totaled about $220 million at the end of 1998. If Mitchell Energy's stock fell below $8.50 a share, the banks could seize George's shares to cover the loans, and he would lose control of the company. Spiros Vassilakis, who managed George's private investments and the family's interests, grew increasingly worried. When some of the Mitchell children asked Vassilakis what they should do, he told them to "just pray."[11]

After several months of wrangling, Vassilakis renegotiated the terms that the banks used for determining a margin call. Instead of foreclosing on the stock if the price slipped below $8.50 for a day or two, he convinced them to use a monthly average of closing prices. This gave George and the company some breathing room, and by early 1999, the stock began creeping upward. They had dodged a financial bullet, but it was becoming increasingly clear that a permanent solution needed to be found. Mitchell Energy couldn't continue this way, and neither could George.

CHAPTER 22

Selling Out

Amid escalating concerns about the company's future, George got another shock in 1998—doctors told him he had prostate cancer. The disease was treatable, but it was one more reminder that he was getting older. Walking was increasingly difficult, as his knees had broken down from decades of daily tennis playing. His brother Johnny had died two years earlier at age eighty-two.[1] George was eighty years old, and so much of what he had worked for seemed to be slipping away.

The cancer raised concerns not only about how much longer George could run the company, but about what would happen to it when he couldn't. For most of his career, George had resisted selling assets he accumulated, whether they were leases in North Texas or real estate in Aspen. You never know, he used to say, what might happen. But by the summer of 1999, even George couldn't deny the reality facing Mitchell Energy. It still had great reserves in North Texas, and it was unlocking the potential of the Barnett Shale, but its success with fracking hadn't yet bolstered the company's balance sheet. With its aging managers and his ten children as potential heirs, some within Mitchell Energy worried what would happen if the children decided they wanted to keep the business. With so many siblings, it could be difficult, if not catastrophic, to sort out control if George died while still owning a controlling interest. The children would each inherit equal shares of the stock, creating the possibility for infighting. "The company could not afford for George to die and still own 60 percent of the company," Homer Hershey said. "That would have been a disaster."

George and Cynthia had encouraged their children to pursue

interests they felt passionate about. All had their own careers, and none wanted to take over the company. Of the children, Todd would have been the obvious candidate to run the energy business. He had the background and board position, and many in the company assumed he was being groomed to take over. But Todd, too, had other interests. He was in his early thirties at the time, and serving on the board was more a learning experience than a path to the executive suite. “It was a window into a lot of really smart and successful people,” he said. George never indicated that he was grooming Todd to run the company—or looking at his older brothers Kent, Mark, and Scott, all of whom were involved in real estate. In hindsight, Todd wondered whether George hadn’t wished that one of his sons *had* stepped into his shoes. “If he was a different person and communicated in a different way, he might have wished someone in the family could have taken it and run it,” Todd said. “It’s a regret that either I didn’t draw it out of him or he couldn’t quite communicate it.”

Instead, the only solution was selling Mitchell Energy. The first phase began in July 1999, when the company sold most of its assets outside North Texas.[2] It used the proceeds to step up the developmental drilling program in the Barnett Shale. Bill Stevens was finally winning his argument over how the company should deploy its capital. He knew the Barnett was the only asset that would attract a buyer, but the company would focus on the core Barnett properties that would generate the best return—with the shallower, conventional wells still producing enough to cover the higher costs of fracking. In October, the company hired Goldman Sachs and Chase Manhattan to find a potential buyer for the rest of the business.[3] Duke Energy from North Carolina, Devon Energy of Oklahoma City, and Anadarko Petroleum in Houston all expressed interest.

A “war room” was set up in The Woodlands, where the companies could send representatives to scour the details of Mitchell Energy’s operations, from production history to financial data. After reviewing the records, Duke and Devon passed. Anadarko, which a few years later would move into the first skyscraper in The Woodlands (the new owners didn’t adhere to George’s belief in keeping buildings below the treetops), offered far less than George thought the company was worth. He was frustrated. No one seemed to appreciate the value of the Barnett Shale. The 1990s had been the most difficult time of his professional career, and the new millennium wasn't looking much better.

To add to the insult, the dot-com boom was raging. Companies that had no hard assets, no profits, and in some cases no idea of how they would ever make money were selling for billions of dollars. How could they be worth more than the business he had spent a lifetime building? Meanwhile, a once obscure pipeline company in Houston had built itself into one of the country's biggest energy companies by eschewing some of the industry's longest-held tenets. Enron didn't believe in physical assets like wells and reserves. Its executives mocked Exxon, one of the oldest and most financially successful oil companies in the world, as a dinosaur. Enron was going to change the industry by moving trading operations online and creating new markets for everything from natural gas to advertising space in newspapers. (In a few years, Enron would collapse in a wave of accounting scandals and file bankruptcy. Its two top executives, Ken Lay and Jeff Skilling, would be convicted of fraud in 2006.)

George joked that he needed to find a way to drill for oil on the internet, but he struggled to understand what made this new breed of company so valuable—as did many other businesspeople.[4] He became obsessed with the idea of tapping that value, telling Todd that Mitchell Energy needed to "build a web site."[5]

In Oklahoma City, Larry Nichols, the chief executive of Devon, met with the team that reviewed Mitchell Energy's books. They were unimpressed. The fracking technology was an interesting experiment, but the technique wasn't economically viable, they told him. The Barnett simply wasn't producing enough gas to justify the expense. Based on those findings, Nichols decided that Devon wouldn't make an offer.[6] George Mitchell appeared to be another victim of the natural gas business, just like Boone Pickens before him. "It was a failed auction," Nichols said. "It was over. Mitchell Energy was off the market."

By April 2000, George had had enough. After months of waiting, all he had from the auction process was a single lowball offer. No one was putting a fair value on the Barnett. Seventeen years of investment, of stubborn perseverance, and the reward was a piddling bid from Anadarko.[7] George was still the majority shareholder, and he remained convinced of the Barnett's value. If no one else could see it, then why sell?

George and Bill Stevens developed a new strategy. With no interested buyers, their only choice was to prove that fracking worked. Mitchell Energy would ramp up production from the Barnett Shale, fracking its way to new levels of profitability. Gas prices were finally rebounding, and if would-be suitors couldn't see the value, then George would show them. The company announced that after reviewing its options, it had decided to stay independent. It also said that it expected by year's end to report its highest earnings per share ever—30 percent more than what most analysts predicted—and it would boost natural gas production by 15 percent.[8]

◈

Typically, when companies try to sell themselves and can't find a buyer, things don't end well. They can cut jobs and other expenses, they can file for bankruptcy, or they can sell off assets piecemeal. The employees usually don't wait around to find out. But Mitchell Energy wasn't a typical company. Not only was George making good on his pledge to boost shale production, but few employees quit after the auction. "All of us were kind of relieved [the sale] didn't happen, but we all knew it was just a matter of time," said Herb Magley, the longtime manager who had run the North Texas operation.

For years, George had turned a blind eye to his employees' flaws. If one failed on a particular project, he might say, "He can't be good at everything."[9] George didn't pay his employees more than other oil companies, and he didn't dole out extra perks. Even the Barnett team didn't receive any extra bonus for their success. Nevertheless, George's personality and his vision, which employees had embraced over the years, engendered steadfast support and loyalty—loyalty that remains among many former Mitchell Energy employees and associates to this day. Far from fleeing the company when no one wanted to buy it, many eagerly joined George's efforts to double down on the Barnett. "George is a unique person," Shaker Khayatt said years earlier. "You admire George. You respect George. We live vicariously through George, his vision, his dreams, his accomplishments. Seeing things like The Woodlands, its evolution, it's a unique perspective. Seeing some of the things he's done in the oil and gas business is very rewarding. . . . Most [employees] could have gone to work for Exxon or any other company, but it was the man, not the company, that attracted them."[10] (By the late 1990s, after low oil prices prompted a wave of industry layoffs and mergers, there were fewer jobs at competitors, and almost none for people interested in working on shale projects. Nonetheless, employee loyalty to George was a major reason people stayed at the company.)

Fracking, his greatest achievement, came to fruition only because George was George—visionary, stubborn, relentless, and willing to spend far more than most other companies could ever have justified to their shareholders. No one knows how much Mitchell Energy actually spent to perfect fracking, but Dan Steward, who coordinated the Barnett team for most of the project, estimates that it was at least a quarter of a billion dollars. That figure doesn't include well costs or the underwriting of early fracking experiments by the federal government, only the costs of the fracking itself. If well costs, seismic surveys, and land acquisition are included, Mitchell Energy's investment was probably more than $1 billion, he estimates.

Bill Stevens, meanwhile, was still as focused on the facts as ever. He understood that the Barnett Shale was a huge resource, but for years he questioned whether it could be developed economically, and more specifically, whether expanding Barnett production made sense. The company had boosted Barnett output by 24 percent in 1998, but earnings continued to slide, and in 1999, Mitchell Energy reported a $56 million loss. In 2000, after the company had doubled down on the Barnett, its fortunes began to improve. It drilled eighty-five new gas wells in North Texas, and natural gas prices soared for most of the year to $9.88 per thousand cubic feet from $2.18. The higher gas prices more than covered the cost of the Barnett wells, driving the company's profit for the year to $175 million.[11] Mitchell Energy's stock price rocketed from twenty-one dollars a share to more than sixty-three dollars by year's end.

Bill Stevens now saw what Mitchell Energy needed to do, and

George's declining health only made the choice clearer. Fracking might not save the company forever, but it just might make it attractive enough for potential buyers to take another look.

◈

Sitting in his office in Oklahoma City in mid-2001, Larry Nichols thought something was amiss. He was reading Mitchell Energy's annual report from the previous year, which it had recently filed with the Securities and Exchange Commission. Production was rising sharply—a 20 percent increase from a year earlier—and costs were going down. How could that be? His team had reviewed the data just the previous year and advised him against making an offer. They described a company that owned one big concentration of natural gas reserves that were playing out, an aging management that seemed to lack the interest or the ability to reinvent the business, and a gee-whiz technology that, while intriguing, didn't appear to be generating results. Had they missed something?

Nichols checked with the team again. They assured him that what Mitchell Energy was doing with slick-water fracking didn't work economically. Nichols had faith in his engineers. A former lawyer, he had founded Devon with his father in 1971, and in the 1990s the company had grown rapidly by buying competitors from Houston to Calgary. Just a year earlier, Devon had acquired Santa Fe Snyder for $2.5 billion—a much bigger target than Mitchell Energy.[12] His people knew how to evaluate companies, and they knew how to put deals together. Devon's track record proved they were good at it.

Nichols studied the numbers again. What had they missed? There was only one way to find out. "I called George and I said, 'we'd like to have another look,'" Nichols recalled. "And he said, 'sure, come on down.'" Once again, Nichols dispatched an engineering team from Devon's headquarters in Oklahoma City to The Woodlands. Once again, they scoured the company's finances and production reports. A couple of days later, they returned and told Nichols that there might be something to the fracking strategy after all.

By the time the Devon team arrived, Mitchell Energy had drilled more than seven hundred vertical wells into the Barnett, mostly concentrated in one area of the field as Stevens had proposed. But something had changed in a year and a half. "They were getting commercial results with that fracking technology," said Tony Vaughn, a member of the team who's now Devon's chief operating officer. "We were seeing more repeatable results."

If fracking worked, it meant Mitchell might be the classic definition of a diamond in the rough. The Devon team determined, as Nick Steinsberger had five years earlier, that slick-water fracks had promise.

As they came to believe in the possibilities of the Barnett, the Devon team also stumbled on something else that could add to shale's potential. In one corner of Wise County, Mitchell Energy a few years earlier had applied a technique called horizontal drilling. It was similar to the old slant-hole technique the company had used on the Lafitte's Gold project in Galveston decades earlier, but the technology had improved. Instead of simply angling the drill bit in a specific direction, the drill bit turned as it descended into the Barnett, following the undulations of the shale horizontally. Once it was in the shale, the well was fracked at multiple intervals along the horizontal portion.

Vertical wells produced limited results because they were limited to the vertical expanse of the shale formation. It was like poking a fork into a layer cake—the prongs pierced each layer from top to bottom. If you wanted to get frosting from the middle layer, you were limited to the depth of the band of frosting. But if you slid the fork in from the side of the cake, it would follow the line of frosting all the way across. Drilling horizontally along the formation would allow for contact with many more fractures, potentially producing substantially more gas per well.

Mitchell Energy had drilled only three wells horizontally, and while the technology worked, the results weren't significantly better than for the vertical wells. Nick Steinsberger had referred to them as an engineering success and an economic failure.

Bill Stevens quickly shut down the horizontal program. There would be no years-long experiment like what Steinsberger had done with slick water. At a meeting of geologists to discuss the Barnett,

Stevens declared, "I'll be dead before we drill any more horizontal wells in the Barnett." After he left the room, Kent Bowker turned to one of the other team members. "Now we gotta kill Stevens," he quipped. This time, though, Stevens won out.

Stevens had a different goal. Combining horizontal drilling with fracking offered the potential to unlock the huge gas reserves of the Barnett to a far greater degree than vertical wells could. It was that potential that Stevens wanted to preserve, in hopes of using it to attract a potential buyer. Drilling more horizontal wells might improve the company's production, but it might also have the opposite effect. What if the technology failed? What if the results were mediocre? Instead, horizontal drilling offered an enticing prospect to would-be suitors. In the sexist terms of Wall Street deal making, Stevens used the allure of horizontal drilling to show a little leg. He also realized they needed to act quickly. The surge in gas prices that had made the company's 2000 profit so strong had quickly faded. By the summer of 2001, gas had fallen below four dollars per thousand cubic feet, almost half what it had been at the start of the year.

The strategy worked. Devon's engineers believed the company could use the horizontal techniques to frack a much larger area of the Barnett. This dramatically changed the economics of the merger. "From the first time we looked at Mitchell Energy to the second, the difference in value was the Barnett," Vaughn said. "Without the Barnett, that company was really not much."

Within weeks, Devon was ready to make an offer. Nichols negotiated across the table from Stevens, with George monitoring the discussions. As the largest shareholder, he had the final say.

The transaction came together quickly, Nichols recalled. "Their lawyers and ours knew where the middle of the road was, knew what was acceptable, and didn't waste a lot of time," he said. The Mitchell team didn't have a lot of options. Despite the spike in Mitchell Energy's 2000 production numbers, the chances of another buyer emerging were unlikely.

In the end, Devon agreed to pay $3.1 billion in cash and stock—equal to more than sixty dollars a share—some 30 percent above Mitchell Energy's share price at the time. Devon also agreed to take on about $400 million in Mitchell Energy debt. As the biggest shareholder, George reaped most of the proceeds, but he didn't take it all in cash. Instead, he accepted about $1 billion worth of Devon stock, which would allow him to benefit if fracking proved to be the success he thought it would.[13] George became Devon's biggest shareholder, and that influence enabled him to impose another condition: he wanted to control a board seat, and he wanted his son Todd to sit in it.

Devon agreed to the terms, and with a stroke of the pen, the company that began with the unimaginative name of Oil Drilling Company ceased to exist. Many former Mitchell Energy employees wonder to this day whether George sold too soon. If he had held out longer, the profits from fracking might have enabled him to keep the company in the family. But many of Mitchell Energy's longtime employees and senior executives were in their sixties by then. If the company had stayed independent, it would have faced unprecedented growth and a new direction that probably would have required new leadership.

Devon, on the other hand, had the management team and resources in place to handle the expansion. After the sale, it stepped up the vertical drilling program, pushing the resources of equipment and people to the limit, and it told the former Mitchell Energy employees in the field to drill horizontal wells. "We told them 'look, we don't care if you make good wells or bad wells, just go drill six of them, frack them the way you frack wells today,'" said Bill Penhall, a Devon senior vice president who helped reevaluate Mitchell Energy. Of the first six horizontal wells Devon drilled in late 2002, five produced good enough results that the company kept trying. It decided to drill about fifteen more, and by 2004, it had improved the process. "During that two-year period, we got more proficient," Vaughn said. "We knew we were on the right track."

It didn't hurt that by 2003, gas prices were on the march again, and they weren't going to stop anytime soon. Most of the country wasn't paying attention to what Devon was doing in North Texas. On the Gulf Coast, companies invested billions of dollars in terminals to import natural gas. The conventional wisdom was that gas, like oil, would grow increasingly scarce and more expensive. By the end of 2005, spot market gas prices reached almost $14.50 per thousand cubic feet.[14]

Devon exported the fracking technology to other basins around the country where it believed the shale was similar to the Barnett, and it replicated its success. Other companies took notice, and now they, too, began fracking shale formations in Arkansas, Pennsylvania, New York, and Ohio.

Bill Stevens left the company when the deal closed, and in 2004 he was named to the board of EOG Resources, a Houston-based company that had once been the oil and gas division of Enron.[15] Enron had sold it years before its own infamous collapse, and EOG had thrived as a domestic oil producer. Now, having watched the ongoing developments in the Barnett, it wanted to take the technology a step further and use fracking to extract oil from shale. Many of the true believers on Mitchell Energy's old Barnett team—most of whom left the company after the sale—didn't believe fracking would work with oil. Oil molecules are bigger than those of gas, and the shale simply wasn't porous enough to allow the oil to flow out, they reasoned. But once again, fracking defied the doubters. EOG began producing oil from shale formations in South Texas, and the state found itself on the cusp of another boom. Soon, companies were using it to squeeze oil from shale formations in places like North Dakota, which had never been a hotbed for energy production before.

George, of course, also left the company after the deal closed, but not surprisingly, he didn't walk away. The stock position meant that he was keenly interested in Devon's success—something that Nichols soon became aware of. Just as Bill Stevens had found out, employees' loyalty to George ran deep, and old habits didn't die with the merger. For years, George had monitored the data on every well his company drilled, and he kept in touch with former employees in the field. "He knew all the wells intimately," Nichols said. "I would periodically get a call from George" pointing out potential problems with a particular well. "He'd say 'you better go check it out.'" Devon, of course, had thousands of wells, and unlike George, Nichols didn't study the data on each one as it came in. But he would pass the message on to the operations people. "For several years, he kept pretty good track by calling field hands and finding out what was going on," Nichols said.

Now that George was the company's biggest shareholder, he also kept a close eye on the company's dividend, which Devon typically declared in March. In late January or February, George would call Nichols and remind him of the importance of dividends for shareholders and urge him to increase them. "You would think I was talking to a poor pensioner who was barely making ends meet and this dividend was highly important to his lifestyle," Nichols said, chuckling. "We just gave him a billion dollars in cash and a billion dollars in Devon stock."

By 2007, the shale revolution was in full boom. About eight thousand wells had been completed in the Barnett, which was being called a second Spindletop, and George was being dubbed the "father of fracking." Vaughn attended a luncheon in Houston and found himself seated next to George. Vaughn introduced himself, thinking perhaps he might enjoy some casual conversation with the man responsible for so much of his company's success. He didn't know about George's aversion to small talk. No sooner had George heard what Vaughn did than he began peppering him with questions about individual Barnett wells. He remembered well names, production details, and drilling plans for each well. "Did you guys ever go back and drill that well?" he asked. It was as if he were still in charge, calling his geologists on a Sunday night. A decade later, he remembered every detail. Vaughn began to understand how Mitchell Energy had succeeded in making fracking work when companies such as Chevron hadn't. "You can't overlook the work that Mr. Mitchell did—to have the fortitude to drill the sheer volume of wells that they did," Vaughn said. "It completely changed the trajectory of the oil and gas industry in North America." In 2007, President George W. Bush declared that America was addicted to oil. Shortages drove prices to $147 a barrel. Less than a decade later, average US gasoline prices were at ten-year lows, crude oil was almost $100 a barrel cheaper, US oil production was at a forty-year high, and the country was exporting crude for the first time since the Arab oil embargo of 1973.

"You have to trace all of that back to Mr. Mitchell's fortitude to keep pestering the Barnett until it worked," Vaughn said.

CHAPTER 23

Attempted Retirement

After the Mitchell Energy sale, George faced a life unlike any he'd ever known. For one thing, he had money. "It was the first time in his life that he had cash," his son Scott said. "He was 83 years old and he actually had cash. I'm not sure he knew what to do with it. He should have gotten an honorary PhD in leverage. He banked everything he put his fingers on. He built mountains of debt." Now, he had more than a billion dollars to do with what he pleased. His kids didn't need the money, and for George, the answer was obvious: give it all away. "He is a funny fellow. He can be the most generous of men yet the toughest in business negotiations," longtime Mitchell Energy board member Shaker Khayatt once said.[1]

George, of course, didn't ease into retirement gracefully. He still worked the phones, talking to former employees about what was happening with his old company's wells and badgering Larry Nichols about dividends. But selling the energy company didn't come with any of the sadness George had over selling The Woodlands. "I don't recall a lot of talk about it," said his daughter Sheridan Lorenz, who spent a lot of time with her father that year. "I think he was kind of relieved."

The expansion of fracking was beginning to take hold, and the story of how George had relentlessly pursued the technology for seventeen years was already becoming the stuff of industry legend. Even if he wanted to, the "father of fracking" couldn't deny the siren call of shale. For all the hard work he'd overseen, George hadn't really reaped the full reward of the technology he and his team had developed. It was Devon that earned the economic benefit of the

Barnett by pairing fracking with horizontal drilling. Sure, George profited from Devon's rising stock price, but at his age, what did he need more money for? What he enjoyed was the thrill of discovery, of putting deals together and watching them grow. George couldn't simply leave all that behind. He never had use for leisure time, and his knees had gotten so bad he could barely play tennis anymore. Even walking had become difficult, and he usually relied on two canes to get around. For someone more interested in intellectual pursuit than relaxation, retirement meant something different—a chance to do it all again.

George got that opportunity thanks to the connections of his son Todd, who would help reunite key players from the Mitchell Energy shale team.

◈

Todd Mitchell had decided he wasn't going to approach his father about his latest business venture. All the Mitchell children were successful, and because there were ten siblings, they had an unspoken understanding: they didn't want to set off a free-for-all by clamoring for their father's money. But the more Todd thought about it, the more he realized it didn't make sense not to show a shale deal to his dad.

Todd had worked the most closely with George over the years. Although he was never a Mitchell Energy employee, he had served as the de facto liaison between his father's determined visions and Bill Stevens's stern pragmatism. During much of that time, Todd had his own oil and gas interests. He earned a master's degree in geology from the University of Texas, where he met Joe Greenberg when they worked together on a research project. They would become lifelong friends.[2]

After they graduated, Todd worked for George's friend and former employee Howard Kiatta, and Greenberg joined Shell, where the company put him on a fast-track management training program. In a few years, both men struck out on their own, and in 1999, Greenberg formed a company, Alta Resources, and convinced Todd to become a passive investor. Greenberg's friendship with Todd gave him a unique perspective on the developments in the Barnett Shale. He knew that George had produced gas from the shale, and that the effort wasn't economical.

As prices began rising in 2000, though, and the pundits predicted gas might become as scarce as oil, Greenberg realized that Mitchell Energy was sitting on the solution that would avert the next energy crisis before it happened. It wasn't long after George sold his company to Devon that the Barnett became one of the country's hottest resource plays.

By then, a tiny business like Alta couldn't afford to buy leases in the Barnett. Thanks to Mitchell Energy's success, it seemed as if every land man in Texas was flocking to North Texas. Greenberg had another idea. George hadn't unlocked gas just in the Barnett; he had developed the technology that would potentially produce gas from *many* shale formations. If the Barnett was number one, what was number two?

Greenberg became convinced that the answer was a formation that arcs across the northern part of Arkansas known as the Fayetteville Shale. Geologically, it was similar to the Barnett, and Greenberg believed that using the combination of slick-water fracking and horizontal drilling, Alta could unlock gas from the formation just as Mitchell Energy did with the Barnett.[3] The opportunity was huge. Todd realized that his father's expertise, as well as his money, could help get the company off the ground. He set up a meeting with George and Greenberg, and for an hour and a half, George peppered Greenberg with questions, interrupting the presentation and poking holes in the pitch. He was back in his element. By the end, he agreed to put up a quarter of the money Alta needed. "He was a tough guy to show a deal to, because he would interrupt you, and sometimes he seemed not to listen," Todd recalled.

Another Houston oil company, Southwestern Energy, had already bought a lot of Fayetteville acreage, but Greenberg found a large parcel for twenty-five dollars an acre several miles to the northeast of the other company's land. Greenberg studied the science, just as George had done decades earlier in looking at the Frustration Fields, and he staked out an area away from the established production. He had hired a geologist, Richard Grigsby, who mapped the Fayetteville and identified "sweet spots" that had been overlooked. As soon as Alta started buying, lease rates shot up. It wasn't long before the price had quadrupled to one hundred dollars an acre.

Greenberg estimated that drilling the first well would cost about $2 million. Because the company was small, it hired another firm to do the work and oversee the production. In the meeting, George warned Greenberg that the most important thing for ensuring drilling success was hiring the right crew and using the proper equipment. Greenberg humored him. He'd known George for years through Todd, and he certainly knew how to drill a gas well.

But the project quickly ran into problems because of the very things George had warned about. For someone who didn't like to deal with details himself, George understood how quickly not paying attention to them could undermine a project. The rig developed mechanical problems. Then, when the well was finally completed, the company hired to pump cement down the well bore to seal it cemented the entire well by mistake. "It was basically this big piece of cement underground," Greenberg said, shaking his head. By then, the drilling costs had escalated to more than $8 million. Greenberg was boarding a plane with his wife for a long overdue vacation when he got a call from George wanting an update on the project. He had to tell him that his investment was now little more than a buried chunk of concrete. Greenberg explained the situation and assured George it would never happen again. They would do things differently next time. From now on Alta would operate its own wells, and it would make sure it had the right crews and equipment. But, he insisted, the assets were still good. He was convinced that slick-water fracks would work in the Fayetteville. There was a long, awkward pause. "OK, mighty fine," George snapped, and hung up. Greenberg sighed in relief.

"He stuck with us, and it was a real testament to both the kind of guy he was as an investor and also his vision," Greenberg said. "He had the vision to see that eventually we would figure it out."

Just as George never criticized his engineers for dry holes at Mitchell Energy, he didn't get angry or blame Greenberg for what had happened. And just as he had on the Barnett project, he had confidence that Alta would solve the problems of the Fayetteville. To improve Alta's oversight, Greenberg hired Jennifer McCarthy as chief operating officer. She had earned a petroleum engineering degree from Texas A&M, winning the watch that George still donated every year to the top student. She made sure to wear it the first time she met him.[4]

McCarthy wasn't the only new hire that Greenberg brought in to spruce up Alta's expertise. Working with Todd, he tracked down Dan Steward, who mentioned Nick Steinsberger as the engineer who had perfected the slick-water fracking process for Mitchell Energy. When Todd told George they wanted to bring Steinsberger on board as a fracking consultant, George sang his praises. He may never have told Steinsberger directly, but he respected his role in unlocking the shale.[5]

Under McCarthy's and Steinsberger's guidance, the company drilled twenty successful wells in the Fayetteville and partnered with Southwestern Energy on about eighty more. The shale boom was gaining steam, and other companies flocked to Arkansas looking for acreage. Alta and Southwestern already held the best leases, and dozens of other companies wanted what they had. Greenberg started getting offers that were far higher than anything he expected. Alta had invested about $300 an acre in the Fayetteville and another $1,200 for drilling costs—or a total of about $1,500 an acre. Other companies offered him more than $10,000. He called George and told him they should sell.[6]

"Nope," George said. "Bad idea. You should grow a company around this." For George, it was a strategy that had served him for more than fifty years and made him a billionaire. He held on to his leases for decades, frustrating many landowners in North Texas and elsewhere. He kept them as prices rose, and he refused to sell as they fell again, doggedly trying to unlock the Barnett. You didn't give up leases, because you never knew what might happen or how valuable they might be one day. Their future potential held the greatest value, at least to George. As he saw it, the stronger the price being offered, the more it proved that once the leases were drilled, they'd be worth even more. "He wasn't big on the net present value analysis," George's son Scott said. "That wasn't his deal."[7]

George's advice to Greenberg was simple—hold assets, build a company, take it public. Greenberg saw the situation differently. "My perception was that we had done almost the equivalent of building a Macintosh in a garage and we should realize some of the value because the appreciation was so large," Greenberg said. George had no intention of selling his interest, but he wouldn't stop Greenberg from entertaining offers from potential buyers. Within a week, Greenberg had a firm bid for more than $13,000 an acre—nine times Alta's investment. He told George he thought they should take the deal. Perhaps the firmness of the offer convinced George, or perhaps he didn't want to interfere with Greenberg's decision. As usual, George didn't share any insights into his thinking. "OK," he said. "Mighty fine." "We had a long conversation about why we shouldn't sell and a very short conversation about why we should sell at the price we got," Greenberg said.

The deal seemed to reinvigorate George. He hadn't been involved in the day-to-day operations at Mitchell Energy in its final years. With Alta, he found himself back in the business he loved. "You could see the wildcatter in him," McCarthy said. "He couldn't get rid of that."

While the Alta project was rekindling George's love for oil and gas deals, it also helped renew his relationship with Todd. During his years on Mitchell Energy's board, Todd was the only one of his siblings still living in Houston. He and his wife, Wendy, would frequently visit George and Cynthia on the weekends or meet them for dinner. Wendy owned Mission Burritos, a well-known Houston restaurant, and George loved talking with her about the eatery. "He liked to hear about the business and he would ask her questions, just like he did in business meetings," Todd said. "It gave him a way into the conversation since he didn't do small talk." Wendy felt a deep affection for her father-in-law as well. After the Mitchell Energy sale, Todd decided to focus on renewable energy and leave Houston—moving first to Scotland and then to Aspen, where he had spent so many Christmases growing up.[8] Todd felt bad about moving when he did. George had suffered through septicemia in 2001, and Cynthia was showing signs of memory problems. The Alta deal helped Todd maintain the connection with his father through regular meetings and phone calls. Greenberg filled in as Todd's local proxy. George called Greenberg frequently to ask questions about the project.[9]

Before Alta sold the Fayetteville interests, Grigsby had perfected a mapping algorithm that enabled the company to zero in on the best places to frack. Now, flush with cash from the sale, Alta applied Grigsby's algorithm to the next promising formation, the Marcellus Shale, which runs from northern Pennsylvania to upstate New York. Houston-based Cabot Oil & Gas had drilled a few vertical wells, but no one had done any horizontal drilling. Alta leased properties for $1,000 an acre, and prices quickly jumped as high as $1,700. The Marcellus had the best geology Greenberg and his team had seen outside the Barnett—even better than the Fayetteville. "We decided we would rather pay a little bit more and stay in this geology," he said.

George, however, disagreed. "That's too much," he responded. "We leased for $100 an acre in Arkansas. I'm not paying $1,700 an acre."

Greenberg replied, "Well, Mr. Mitchell, we just sold Arkansas for $13,000 an acre, and this looks better—this looks way better."

"Well," George said, "I'll go in for a little bit."

George understood that the success of the Barnett and the Fayetteville was driving up lease rates. After all, he had gobbled up thousands of acres in the Barnett in the 1950s to get the jump on just this sort of land rush. But he had also seen dozens of oil companies go bankrupt because they got swept up in the hype and overpaid for leases. His reluctance was his way of reminding Greenberg to keep a careful eye on costs.

In 2008 and 2009, in the wake of the global financial crisis, banks tightened lending requirements, leaving many drillers strapped for capital. Backed by George and several other wealthy investors, and with its coffers flush from the Fayetteville sale, Alta used the crisis to its advantage. It snapped up fifty thousand prime acres in the Marcellus. With every ten thousand acres Alta leased, Greenberg would go to George and ask whether he wanted to buy a little bit more. The verbal dance was always the same.

"We're now paying $2,000 an acre," Greenberg would say.

"You shouldn't pay $2,000 an acre," George would reply. "That's too much."

Greenberg would point out how much they had sold acreage for in the Fayetteville, and how Cabot had drilled some promising wells in the area.

"Ah, I'll go in for the next ten thousand," George would say.

Greenberg and McCarthy knew that George would ultimately stay in the play, but they also knew that he wanted them to be cautious. "He didn't want to just say yes," McCarthy said. "He wanted to ask the questions and make sure we had the right answers."

After they acquired the acreage, another partner came to Greenberg with an analysis that showed that the fracking boom would cause natural gas prices to collapse from oversupply. He recommended that Alta sell its Marcellus holdings—and soon. Greenberg and McCarthy studied the report and decided it was accurate. Fracking for natural gas was about to become a victim of its own success. Once again, Greenberg went to George and said it was time to sell. And once again, George told him that was a mistake. The Marcellus was likely to be the most prolific shale gas play in the country, if not the world—bigger, even, than the Barnett. Alta should stay in, develop the field, and go public. "He was pretty adamant about it," Greenberg said. Once again, Greenberg fielded some offers, and about a week later, he called George. The company had a deal for $14,000 an acre. Just as with the Fayetteville, the conversation on the sale was short—George agreed instantly. Alta had invested about $170 million in its Marcellus holdings and sold them a few years later for $690 million. One month after the sale, natural gas prices, which had hit $6.50 per thousand cubic feet, began to fall. They eventually dropped as low as $1.80.[10]

After the sale closed, Alta threw a party at the Flying Saucer, a downtown Houston pub known for its selection of beer. George was almost ninety and used an electric scooter to get around, but he came to the party to have a drink with the Alta team. He still loved the thrill of deal making and the camaraderie of a company.[11]

Alta began looking for its next deal in 2010. But times had changed. Natural gas prices were falling, and companies like Chesapeake Energy, which had quickly become one of the country's biggest natural gas producers, were paying premiums for leases. At the same time, fracking had unleashed an abundance of gas. Alta faced a daunting prospect: higher costs and lower revenue. Greenberg turned his sights to Canada and bought properties in an Alberta shale formation known as the Duvernay. This time, George didn't quibble much about the acreage price, which was $3,000. "Mr. Mitchell," Greenberg told him, "we're going to Canada because gas prices in the US are depressed." George smiled and said, "Yeah, I ruined it for everybody."

CHAPTER 24

Wendy's Telescope

Robert Gates walked into the conference room in George Mitchell's office on the thirty-sixth floor of Chase Tower in downtown Houston with its sweeping views of the city. He wasn't there to admire the scenery. Gates would later become George W. Bush's defense secretary, but in mid-August 2005, he was president of Texas A&M, George's alma mater. George had proposed a major contribution, as much as $10 million, to fund an institute for fundamental physics. Gates had agreed that the university would put up the remaining $2 million of the estimated costs.

As Gates, university officials, and various attorneys settled around the conference table, Gates opened the conversation by saying that during his tenure as university president, "the physics department has not been one of my priorities." Before Gates could finish the thought, George jumped in. "Yes, I've noticed," he quipped.

It really hadn't been a priority of George's, either, until about the time he was selling Mitchell Energy in 2001. George had never forgotten his frustration over the cancellation of the Superconducting Super Collider. One night, he saw a television science special on the local Public Broadcasting Service channel that featured an interview with Stephen Hawking, the famous theoretical physicist. Asked about his greatest disappointment, Hawking said it was the US government's failure to complete the Super Collider. For years, Hawking criticized the decision at every chance, calling it "short-sighted."[1] His comments resonated with George. George had long admired Hawking for his groundbreaking work in cosmology despite his physical impairment of amyotrophic lateral sclerosis, or Lou Gehrig's disease, which had left him wheelchair bound and speaking through a voice simulator. Now, George began following Hawking more closely. Like many chief executives, he had a clipping service that collected articles of interest, and he frequently added topics he wanted saved—items that ranged far beyond oil and gas or real estate. He added Hawking's name to the list, and he was soon reading almost every article in which Hawking was quoted.

If George didn't understand a quote, he would call Peter McIntyre at Texas A&M. "I'd get these phone calls about every two weeks from George: 'Peter, Hawking said so and so the other day. What does that mean?' So, I would pull it up on my computer." Cosmology wasn't his field, but McIntyre would try to explain Hawking's comments. By then, George was preoccupied with saving Mitchell Energy and making fracking work, but he kept following Hawking. At the end of one call, he told McIntyre he wanted to bring Hawking to Texas, but he didn't know how. "If you think of a way to do it, let me know," he said.

Reading about Hawking reawakened an interest in astronomy and the universe from George's childhood. Through the years, he had liked looking at the stars, but the more he talked with McIntyre about Hawking's comments, the more his passion grew. "When I was just getting out of high school, I was very interested in cosmology," George recalled in 2009. "I read all that I could find out about it and even strongly considered becoming a physicist." The consideration was derailed when he went to work with Johnny in the oil fields of Louisiana and decided to study petroleum engineering. "I decided I had better go into something where I could make some money," he said.[2]

Now that he'd made his money, George could indulge his latent curiosity. At A&M, his renewed interest hadn't gone unnoticed. When Ed Fry took over the physics department in 2002, it was struggling financially, and Fry needed to raise money. McIntyre mentioned George's fascination with Hawking, which gave Fry an idea: Would George be willing to pay to bring the famous physicist to A&M for a guest lecture? McIntyre agreed to ask, but first, he wanted to make sure Hawking was willing to make the trip.[3]

He approached a fellow professor, Christopher Pope, who was

an expert in the theoretical framework of physics known as string theory. Pope was the only string theorist on A&M's faculty. Another one had been lured away by another university, and rumors were swirling on campus that Pope was mulling other offers as well. Pope had been Hawking's student at Cambridge University from 1976 to 1979. He had developed a close relationship with the famous physicist, often carrying Hawking's wheelchair up and down stairs at the 700-year-old school.[4]

If anyone could convince Hawking to come to Texas, it was Pope. And if anyone could convince George to fund a chair that would keep Pope at A&M, it was Hawking. A&M's development foundation, which coordinated funding efforts with wealthy donors, had tried in vain for years to get George to make a significant donation. He gave most of the $4.2 million for the new campus tennis center, which opened in 1998, and he had provided one hundred acres on Pelican Island for A&M's Galveston campus thirty years before that, but he had never been interested in making a major donation in his chosen field.[5] The geosciences building was named after fellow wildcatter Michel Halbouty, with whom George regularly had lunch in Houston. George argued that petroleum engineering and geology had plenty of wealthy alums to give them money. "They put these beautiful plans together for something *they* wanted, and George just wished them good luck," Fry said. "I went to George with something George was interested in, which was Stephen Hawking."

Pope agreed to go to England and discuss the idea with his mentor. "I explained the whole thing to him, and he was quite enthusiastic," Pope said. Pope returned to Texas with the good news. Hawking, who had spent his entire career in academia, understood the importance of large donors to research programs. He was planning a trip to the California Institute of Technology a few weeks later and suggested the group meet there and discuss the Texas trip in more detail.[6]

George, Pope, McIntyre, and Fry took a commercial flight to Los Angeles and then drove to Pasadena, where Hawking rented a house. The professors noticed that almost from the moment they met, George and Hawking had an unusual rapport, kindled by their mutual disappointment with the canceled Super Collider. "George

Mitchell really felt very close empathy with Stephen on that point," Pope said. Their shared remorse formed the grounds of a lasting friendship. Hawking not only agreed to come to A&M and stay for a month, but also suggested that the university establish an institute for fundamental physics.[7]

A few weeks after they returned to Texas, Fry, Pope, and McIntyre drove to The Woodlands for breakfast with George. They planned to ask him for $5 million. That would endow a chair in physics for Pope and pay the cost of bringing Hawking to A&M for a month of guest lectures. George listened politely as the professors made their pitch, and then he leaned back in his chair and gave a slight chuckle. "I'll give you $800,000," he said, "and that's a helluva lot more than you need to bring Hawking to A&M."[8]

During his visit, George entertained Hawking at Cook's Branch, the Mitchell family's 5,600-acre nature preserve west of The Woodlands. The two spent hours enjoying the scenery and occasionally discussing the importance of cosmology research. "It was rather remarkable, because Stephen is a man of few words, necessarily. And by the time they were meeting, George Mitchell's hearing wasn't that good. They didn't say a lot to each other," Pope said.

The two men also bonded over their physical limitations. Hawking was wheelchair bound and George, by then, needed his electric scooter to get around. In 2006, George flew to Cambridge with his daughter Sheridan Lorenz to meet with Hawking. Todd Mitchell was living in Scotland at the time, and he and his family came down to join them. As they walked across the Cambridge campus, the family lagged behind, allowing George in his scooter and Hawking in his electric wheelchair to go ahead. Todd, in particular, found himself struck by the magnitude of the moment—a Texas wildcatter rolling across the grounds of one of the world's oldest universities talking with the world's preeminent physicist.

During the Cook's Branch meeting, Hawking and George again talked about a fundamental physics research institute at A&M.[9] Fry, who was known on campus for his tireless fund-raising, continued working on George, too. He found a 2005 article in the journal *Science* about the Kavli Institutes.[10] Fred Kavli had amassed a $600 million real estate fortune, which he used to sponsor a series of institutes for astrophysics, neuroscience, and nanoscience. He started by pledging $7.5 million for a center in theoretical physics at the University of Santa Barbara, and he then created another to study particle astrophysics and cosmology at Stanford. Unlike those who funded many research grants, Kavli didn't care about measurable or short-term results. He wanted the research he sponsored to pursue big, open-ended questions.[11]

Fry thought Kavli's approach might appeal to George because of the long-range nature of the research. He made a copy of the article and sent it to Linda Bomke, George's secretary, who put it in George's daily stack of reading material. A few days later, George called. "This information you sent me on Kavli? We can do better."[12]

He once again summoned McIntyre and Fry to The Woodlands and asked them how to create a fundamental physics institute like the one Hawking suggested. First, they said, they'd have to draw up a plan. "Don't make it a little plan," George told them. "You do these things in a big way. Come up with a big plan."[13]

After he visited College Station to review possible sites within existing buildings, he told Fry, "Those buildings, they don't look good. I don't want a Mitchell Institute stuck in one of those things." What the institute needed, he insisted, was its own building. "He was teaching us to think about it in a big way," McIntyre said.

Working with Pope, they fleshed out an idea for how the institute would be structured. It would focus on astronomy, cosmology, and the origins of the universe. To kick off the effort, George pledged $2.7 million to endow two chairs, one named for Hawking and awarded to Pope, who now had no plans of leaving A&M. The rest of the money would go toward a month-long meeting of the world's top physicists, including Hawking. The conference would include lectures—Hawking's popularity filled a 2,500-seat auditorium—as well as physics workshops at Cook's Branch that focused on fresh insights about the universe's origins.[14] The gathering also served as the inaugural meeting of the George P. and Cynthia W. Mitchell Institute for Fundamental Physics.

George loved hosting the event and spending time with so many big thinkers. The success sharpened his determination to see the institute reach fruition.

Meanwhile, Fry worked up a proposal for a new building, which he estimated would cost $5.2 million based on another building on campus that had just been completed. The plans included a round tower that would adjoin the existing engineering and physics building. George liked the plan and agreed to contribute $5 million, but he wanted the New York–based architect Michael Graves to design it. Cynthia loved Graves's work, and she was still annoyed that George had scrapped her effort to have him design a small theater in The Woodlands years earlier. Graves submitted the blueprint, but it cost $3.5 million more than George had pledged. He agreed to cover the increase.

As the work ground on, the costs crept up to $12.1 million. George had hit his limit. He said he wouldn't give more than $10 million, and if the university wanted the building, it would have to raise the difference. Fry worked his way through a series of meetings with university officials, including Gates. He realized that the institute's striking circular architecture—for which George was paying handsomely—would be lost among the other buildings on campus. He developed plans for a new physics building adjacent to the Mitchell Institute. Both would be built on a prominent tract on the northern edge of the campus. Eventually, Fry, Gates, Joe Newton—who was the dean of the science college—and a few other officials wound up in George's office for the big meeting on August 15, 2005. Newton could feel the twinge of history as they sat around the table. Gates, after all, had come to make a deal that would include one of the largest donations ever given to A&M by a single individual. That donor, however, had a reputation as a shrewd and hard-nosed deal maker.

Gates agreed that the university would contribute $2 million to the Mitchell Institute. But Fry had taken George's request to think big to heart. Did George really want his new institute attached to the old physics building? They could build a new building and add it to the institute. As the men gathered in George's office, they were trying to reach a deal on who would pay another $45 million for the second building.

Gates said that he had identified another $20 million in cash from sources around the university (including, Fry claims, a fund that had accumulated from revenue that Coca-Cola paid the university for placing vending machines on campus). If Mitchell would agree to match the contributions, the university would use his money for the new buildings. Gates had turned George's penchant for matching funds back on him.

George said he had some problems with the proposal, but he would give $3 million a year for five years. Gates countered: "The $20 million really straps me. How about you do 10 years?" George countered with $2 million for ten years. Gates paused. Two and a half million over ten years would split the difference.[15] Both seemed ready to agree when Fry spoke up. "It would really help to do it in five years—how about $5 million for five years?" Fry didn't want to wait ten years for the buildings. The room froze. Both sides had been close to a historic deal. Had Fry overreached?[16]

George thought for a moment. Finally, he said, "Same amount of money, so OK." George had agreed to give $25 million for the physics building and another $10 million for the institute, in addition to the $2.7 million he'd already committed.

"I think we both knew where we were headed with the negotiations, but I had fun dickering with him," Gates said. "Some folks I had talked to were kind of intimidated by him, but I found him pretty easy to deal with. I had the benefit that I had spent a number of years as an intelligence adviser to a strategic arms delegation, so I had actually negotiated with the Soviets."

The deal, though, was just the beginning. The university said it couldn't hire Graves because it had to solicit competitive bids for any architect it hired. So, George hired Graves and donated his services to the school.

In all, the buildings' costs rose to $82.5 million, in part because the institute was increased in size to make room for additional faculty. George contributed about $36 million of that. His contributions didn't end there. In creating the institute, Fry and Newton also wanted to add an astronomy program, which A&M didn't have. Starting one meant convincing the state's higher education board that the program would benefit the state and the university. Fry assembled a panel of top astronomers to gather their ideas for building a first-rate program. George, of course, attended the meeting.

Wendy Freedman, the director of observatories for the Carnegie Institution in Pasadena, California, chaired the gathering. She also led the board of directors for the Giant Magellan Telescope, one of three projects to build next-generation ground-based telescopes that would probe deep space for signs of life.[17] She didn't know who this old bald man in a scooter was, but he kept asking questions, mostly about how to put A&M's astrophysics program on the map and how to build a first-rate department. Freedman wasn't sure what to make of the questioning. Was this guy serious? Finally, she told him that a top-notch program wasn't just about bringing in the best people; it needed access to the best facilities, telescopes, and lab equipment. Currently, Freedman's project, the GMT, and its two rival programs, the European Extremely Large Telescope and the Thirty Meter Telescope, represented the vanguard of telescope technology. The other two wouldn't be operational before the mid-2020s, but the GMT could achieve "first light," as astronomers called it, more quickly.

Freedman's answers only heightened George's interest. A few weeks after she returned to California she got a call from Ed Fry. Would she be willing to come back to Texas to meet with George? He would pay for her trip. The three met for breakfast in The Woodlands. George peppered Freedman with questions about telescopes and the GMT in particular. The project was raising $8 million to build the first of seven giant primary mirrors, each more than eight meters—or twenty-eight feet—in diameter. When completed in 2022, the telescope, which would be placed on an 8,500-foot peak in the Chilean desert, would have ten times the resolution of the Hubble Space Telescope. The mirrors would weigh twelve and a half tons each, be curved to a precise shape, and be polished to within one-millionth of an inch, or one wavelength of light.[18] When fully operational, they would capture light from the far reaches of the universe and reflect it through a series of smaller mirrors into the imaging cameras. The cameras would measure the concentrated light to determine the distance of objects and their composition.[19]

George turned to Fry as they were leaving. "I want to give a million dollars to Texas A&M for Wendy's telescope," he said.

Freedman was surprised by George's offer, but as she got to know him, she realized that he relished the role of the catalyst. By making the initial investment, he hoped to generate momentum, so his donations would attract others to the project. Equally as important, funding the GMT would put A&M's astronomy program on the map. "Texas A&M didn't even have an astronomy program at that time," Freedman said. "I didn't even view them as a potential partner in the GMT."

As always, George's $1 million came with conditions. In this case, his offer was good only if the University of Texas also joined the project and contributed the same amount. The two schools were the oldest state-funded universities in Texas, and they enjoyed a rivalry that spanned more than a century. UT's own efforts to contribute to the telescope had been thwarted by a lack of funds. Now, faced with a challenge from their rival Aggies, the Longhorn astronomers convinced their administrators to join the program.

Ultimately, George gave $25 million to the GMT project in addition to the money he gave A&M for the astronomy program.[20] The first mirror was cast at the University of Arizona in 2005, and George flew out for the unveiling. Because of its size and precision, polishing and finishing the giant mirrors took another six and a half years. Freedman nicknamed the first two George and Cynthia.

George wound up giving far more than he'd planned to the physics and astronomy programs at A&M. Again, as with The Woodlands, his passion dictated his generosity. He was frustrated by the lack of public interest in the pure sciences. Businessmen gave money to business schools, and oilmen gave to petroleum engineering programs, but to George, these were just examples of the myopic thinking of too many corporate leaders. Pure science research focused on the biggest questions of all—where we came from, how we got here, how the universe was born. George was constantly pushing his publicist Dancie Ware to generate more acclaim for the GMT. "We need more science writers writing about science," he would say.[21]

George's combined gifts made the GMT possible and vaulted A&M to a leading position in astronomy in less than a decade. In 2016, nine years after its founding, the program featured eight professors and one lecturer and was developing a doctorate degree. "I could not have the astronomy program that I have today—nothing even close to it—if George Mitchell hadn't helped us when he did," said Nick Suntzeff, the astronomer who leads the university's efforts and holds a chair endowed, in part, by George.

But it's the telescope that Suntzeff says could become the greatest testament to George's philanthropy. The GMT is on track to beat the other giant telescopes to first light and begin scanning deep space for signs of life. "It would be a testament to George Mitchell and the other people who provided funding if we could make a discovery of that magnitude with the telescope that they provided funds for," Suntzeff said. "It would be an appropriate payback to their generosity."

CHAPTER 25

Fishing in the Sunsets of Eternity

George had enjoyed good health for most of his life, but about the time of The Woodlands sale and the mounting financial pressure on the energy company in the late 1990s, that began to change. The decline started with his prostate cancer diagnosis in 1998. He responded well to the treatments, but not long after the energy company sale, George's daughter Meredith found him lying unresponsive on the floor of his Galveston beach house. She called an ambulance, and George, who was then eighty-three, was rushed to the hospital. He had developed septicemia from a bladder infection related to the cancer, forcing him to stay in the hospital's intensive care unit for ten days. A doctor warned family members that septicemia in older men can impair other organs. Often patients don't live more than a few months. The doctor asked whether George's affairs were in order.

George, too, was convinced he would die. He scribbled a two-page letter to Cynthia outlining his wishes. He wanted to be buried at Cook's Branch, the family's retreat near Magnolia, northwest of The Woodlands. He ended the note with an appeal that attempted to cut through the years of conflict between them: "I am hoping you will wish to be buried next to me, because I love you," he wrote.

George recovered, but the experience changed him. He became more reflective, often staring out the window of his office at the bustling city of Houston that stretched to the horizon in every direction. "You'd find him sitting there, deep in thought," Dancie Ware said. George had always been a deep thinker, and he'd always been quiet, but now he became even more so. "He became more introspective," she said.

Cynthia, meanwhile, was facing her own struggle. She had been having memory troubles and often got confused by simple tasks like following a map. The problems grew worse, and in 2002, doctors told her and George that she had Alzheimer's. Cynthia's battle with the disease would go on for years, and it was hard on both George and the ten siblings, but if there was a bright spot, it was that it drew George and Cynthia closer than they had been in years. She wanted to hold his hand or sit next to him, and he insisted on helping to take care of her. One of the children remarked that perhaps she had forgotten why she was mad at him.

"She was really the center of the family, and the family was united around her," said Dr. Claudio Soto, a neurologist with the University of Texas Health Science Center in Houston. After Cynthia's diagnosis, George gave $10 million to establish the school's George P. and Cynthia W. Mitchell Center for Research in Alzheimer's Disease and Related Brain Disorders.

During Cynthia's illness, George took charge of her treatments. He was willing to try new procedures, even if they didn't work, and he was particularly interested in experimental therapeutics, frequently asking Cynthia's doctors about new treatments and clinical trials he'd read about.[1] "I was struck by his constant and unwavering good will and support," said Rachelle Doody, Cynthia's doctor at the Baylor College of Medicine in Houston. George gave millions to the college during Cynthia's illness, in part because he felt it important to support the institution that provided her care, but also in the desperate hope that a cure would emerge in time to save her. "Dad was very naïve about medical research," his daughter Sheridan Lorenz said. "He believed that if he threw enough money at it, they would find the right drug. Or maybe it was just the hope of that." Medical research, much like the development of fracking, is a slow process, marked more by incremental advances than major breakthroughs. In addition to the research at Baylor and the UT Health Science Center, George created the George and Cynthia Mitchell Center for Neurodegenerative Diseases at the University of Texas Medical Branch at Galveston.[2]

George had his own methods of coping with his wife's illness. He still talked frequently with his former employees, and he had learned a few years earlier that Herb Magley's wife also suffered from Alzheimer's. He would call Magley, the former North Texas manager for Mitchell Energy, to recount their experiences.[3] "I was surprised he wanted to talk about it, because Alzheimer's has such a stigma," Magley said. "But he was very open. He said 'I'm here for you.'"

The children set up a rotating schedule in which each would look after George and Cynthia when they could. Sheridan bought and restored a historic home in Galveston so she could spend weekends there with her parents, and she built a cottage behind their home in The Woodlands where the children could stay overnight. As Cynthia's health declined, it became more difficult for her to go to noisy restaurants and other public places, and increasingly, the various siblings would gather for family dinners either in The Woodlands or in Galveston. Despite Cynthia's health, George still insisted on living in The Woodlands during the week and driving her to Galveston on the weekends.

His efforts to fund new research didn't develop any miracle cures for Cynthia. She died in 2009, two days after Christmas, surrounded by her family, at their home in The Woodlands. She was eighty-seven and had battled the disease for more than a decade.[4] Cynthia left behind her own legacy, in addition to the one she shared with George. She underwrote a distinguished author's program at the University of Houston and created the school's Cynthia Woods Mitchell Center for the Arts. She quietly funded far more events that were not publicized, and sent several underprivileged children to college.[5] And, of course, her name adorns the Cynthia Woods Mitchell Pavilion, the prominent outdoor performing arts center in The Woodlands.

Her creative drive rivaled George's intellectual curiosity, and she was known for her sense of style and eye for design. The marriage may have struggled, but the mutual respect remained. Cynthia rejected the traditional trappings of wealth, and her strong sense of altruism influenced George's own compassion and philanthropic spirit. Whatever their differences over the years, George and Cynthia remained committed to each other and to the ideas they shared. "They were two kids growing up poor in the Depression, and it's almost impossible to put them into a modern context of why they did what they did," Todd Mitchell said. "Why did they have ten kids? Why did they stay together? They'd already been through broken families, and they weren't going to break this family, and so they were going to stick it out and raise their kids. And they did."

◈

George was ninety when Cynthia died, and he continued living in The Woodlands and driving his old Cadillac sedan to the office in Houston and to Galveston on the weekends. As his knees grew worse, he refused to use a wheelchair. It was Sheridan who finally convinced him to use the scooter, which in his mind lacked the stigma of a wheelchair. He put an Aggie bumper sticker on the back. As his lack of mobility increased, his children convinced him to move back to Galveston, where they had a special suite outfitted for him in the Tremont House, the hotel he and Cynthia had restored in the Strand.

George focused on the places and avocations he found most interesting—Galveston, the A&M physics program, and his investment in Alta Resources. "All of those things were exciting and were about the future and not about the past, which is kind of how he operated," Todd said.[6] In addition to making his regular weekend rounds—meeting his childhood friends for coffee at Kroger, then going door to door among his buildings in the Strand to ask tenants about their business—he relished spending time with his family, especially his grandchildren. He loved taking them fishing, especially since he could not play tennis. For years, he had used his boat, *Where the Fish Are*, to take family and friends around the jetties of Galveston. He liked to give money to his grandkids when they visited, telling them they should "go recycle it on the Strand."[7]

Once, when Todd came to visit with his son and daughter, George decided spontaneously to take them fishing. He didn't have enough poles, so they went to Walmart to buy gear. When they got to the checkout counter, George suddenly slipped into negotiation mode. As Todd recalled: "We're checking out, and they give us the final price. It was like $73—he'd bought a lot of stuff. And he's like,

'That's too much.' We couldn't tell if he was serious at first. Then he says, 'Tell them that's too much. I don't want to pay that much.' And I'm like, 'Are you kidding me? Dad, you're not going to get a better deal than the Galveston Walmart.' The kid behind the counter was like 23 years old and was just looking at us like, 'What the hell is going on here?'"

Todd's daughter, Clare, remembers the days in the Tremont as special times with her "Poupa," as the grandchildren called him, a variation on the Greek word for grandfather, *papou*. He would give them rides on his scooter through the hotel. The scooter wouldn't fit in the regular hotel elevators, so they designed a special route that went behind the office, through the kitchen, and to the freight elevator at the back of the building. "He had every corner measured out to literally the half inch," Todd said. "He was in his 90s by then, and he would round the corners and you would think he was going to hit something, but he would always weave around things." Often, he had one of the grandchildren riding on the back.

George continued to read extensively on a wide range of subjects. One that increasingly drew his interest was climate change. While he made a lot of money off the rising land values from the Alta properties, he also knew that many of his fellow oilmen could be dismissive of environmental concerns. For decades, he'd been frustrated by what he saw as shortsighted thinkers. He knew that the fracking boom would take an environmental toll. Building on George's discovery, oil companies had expanded fracking nationally. More wells were being drilled in the United States than had been drilled in decades, and fracking was bringing them into new regions of the country, often close to residential neighborhoods. Others saw his lifelong commitments to the competing interests of sustainable development and fossil fuels as a paradox, but George never did. He had built an entire city on the concept of sustainability, but to George, sustainability wasn't quite the same as environmentalism. While the two concepts shared many of the same characteristics, and while George loved nature, he saw development and energy as vital to a sustainable future.

In 2010, documentary filmmaker Josh Fox released *Gasland*, which purported to show the evils of fracking. In its most famous scene, a homeowner in Colorado opens his kitchen faucet and holds a lighter to the stream of water, which erupts in a burst of flame. A company had been fracking nearby, and the film concluded that the drilling was responsible for the burning water. The scene was reminiscent of the homeowner claims in North Texas that Mitchell Energy had battled in the 1990s, but the movie never explored other possible sources of contamination or natural methane seeps into the water supply. The Colorado Oil and Gas Conservation Commission investigated the incident and found no indications that drilling caused the contamination, but it didn't matter.[8] The image was seared in the public's mind.

In other fracking locales, similar problems soon followed. In Dimock, Pennsylvania, the water ran brown, smelled like rotten eggs, and became combustible after a company fracked wells in the area. In that case, the driller was found liable for not properly casing the well as it passed through the water table. This, too, was reminiscent of the North Texas lawsuits against Mitchell Energy. The contamination had nothing to do with the fracking itself but with poor drilling practices, so it could have happened with a conventional well. Still, the issues and the industry's lack of a proactive response left fracking as the convenient villain in the minds of the public.

George knew that in the rush to drill, the industry was getting sloppy. Too many oil companies were willing to cut corners to make a quick buck. In Pennsylvania and New York, which were less accustomed to frequent drilling than Texas, fracking quickly ran afoul of regulators and the public.

In 2012, George teamed up with Michael Bloomberg, then New York City's mayor, to write an opinion essay in the *Washington Post* that attempted to take a middle ground in the fracking debate. While extolling fracking's benefits—lower gasoline prices, job creation, reduced dependence on coal, and greater support for renewable energy—the piece called for "common-sense" regulations. Companies should disclose all chemicals used in fracking, reduce water consumption and dispose of wastewater properly, agree to tighter air pollution controls, and curb the impact on roads and communities, according to the authors. "Safely fracking natural gas can mean healthier communities, a cleaner environment and a reliable domestic energy supply right now," they wrote. The two men never met, but both shared a pragmatism that was sorely lacking between the extremes of the energy business and the environmental community.

In private, George was more blunt. "These damn cowboys will wreck the world in order to get an extra one percent" profit, he told his son-in-law Perry Lorenz. "You got to sit on them."[9]

Once again, George was frustrated with both sides. The industry, shortsighted as ever, didn't recognize—or didn't care about—the backlash it was courting if it didn't embrace the proper safeguards. Environmentalists refused to acknowledge that fracking for natural gas was an essential first step toward a cleaner energy future. George didn't believe fracking was bad, but he also believed it should be done right.

Another issue came into play on which George and environmentalists strongly disagreed, at least initially. When George first heard about climate change, he thought it was nonsense. He wanted to rely on the old geologist's thinking that climates are always changing. He discussed the issue with Todd, who told him the science showed that climate change was real. George did more research and talked to several experts. He eventually came to understand their concern. In his final years, he began focusing on climate change as another threat that both the energy industry and sustainable development needed to address. Even at ninety, George was open to new ideas. "You expect a cranky old fossil fuel executive ranting about government regulation, which is what you get from the industry," Todd said. "I loved my dad for his stubbornness, and yet in the middle of that stubbornness, he would reach out and say 'I need someone to explain this to me.'"

In 1978, George and Cynthia had established a charitable family foundation, which was run for nineteen years by their second-oldest daughter, Meredith Dreiss, and later by their granddaughter Katherine Lorenz. Over the years, George and Cynthia discussed their goals for charitable giving with their children. Cynthia was interested in causes such as poverty, education, and social and racial equality. George, meanwhile, wanted to make sure sustainability played a large role in the foundation's activities. He insisted until his final months that part of the foundation's work focus on making grants for programs that research the effects of climate change as part of its broader sustainability efforts. Two years before he died, George also joined Warren Buffett's Giving Pledge, in which wealthy individuals commit to giving at least half their fortunes to charity.[10]

⟡

By 2013, George's physical health had declined significantly. In April, he suffered an ocular occlusion, a stroke in the arteries to his eyes that left him blind. A few months later, after more than a decade of battling prostate cancer, his prostate-specific antigen levels spiked. PSAs are proteins produced by cells of the prostate, and his elevated levels were a sign the cancer was advancing. Old friends and former employees visited, and George seemed to enjoy reminiscing with them—one of the few times he was willing to look backward rather than forward.

George's children knew their father's health was deteriorating, and they wanted him to receive the recognition they felt he deserved for fracking. By that time, a study had found that every American household had saved an average of $1,200 a year because of lower gasoline prices from fracking for oil. President Obama was routinely giving speeches touting the country's new energy independence, and he even noted the government's role in funding the early research. (He also dubiously claimed fracking wouldn't have been possible without the government's funding. While Dan Steward is quick to give the government credit for its role, he notes that "George Mitchell could have developed fracking without the government; the government could not have developed fracking without George Mitchell.") Yet for the man who made it all possible, there was nary a mention outside the oil industry or Houston.

Greg Mitchell, George's fourth-youngest son, who is a biologist and oceanographer, began the process of getting his father nominated for the Presidential Medal of Freedom, which is awarded to people who make important contributions to American national interests, culture, or other significant public and private endeavors.

Greg and his siblings asked an array of influential people they knew—including Stephen Hawking, who had received the award in 2009, former US senators George J. Mitchell and Kay Bailey Hutchison, Robert Gates, and the oil expert and Pulitzer Prize–winning author Daniel Yergin—to write Obama on George's behalf.

As her father's health continued to decline, Sheridan in particular grew increasingly frustrated with the lack of response. Late at night, from her room in Galveston, she wrote her own letters to Obama, imploring him to recognize her father before it was too late. The White House never responded. "It was too political to give a Medal of Freedom award to a guy who'd invented fracking, even though Obama, who I'm a fan of, was saying 'we have 100 years of energy independence,'" she said. In her mind, George was denied the recognition he deserved because of politics.

Because of his elevated PSA levels, the doctors increased George's medications, which gave him hallucinations in his final days. His daughter Meredith was with him, remembering the years when he would take them fishing as children. She said, "Hey Dad, I want to go fishing. Can we go fishing with you?" George nodded his head, and Meredith climbed into bed beside him, and they pretended they were once again on *Where the Fish Are,* heading for the jetties. Perhaps he even imagined one of his favorite scenes: the sun slowly setting over the island of his birth. He died on July 26, 2013, at age ninety-four.

His children chartered a boat and scattered some of his ashes in Redfish Hole, his favorite fishing spot off Galveston where he had spent hours patiently untangling their lines when they were young. Later, they put the remainder of his ashes at the base of a twin pine at Cook's Branch, the family retreat, next to Cynthia's, as he had hoped. Their graves are marked with simple, uncut natural stones, each engraved with their signature. The site overlooks the coastal field, which George thought was the most peaceful place in the preserve. He used to go to the second floor of their house there and stare out at the serenity of the small lake, its glass-like surface reflecting the waning sun.[11]

◈

The family held a private memorial service at an Episcopal church in Galveston, followed by a public celebration of George's life. Two days later, a bigger memorial was held at the Cynthia Woods Mitchell Pavilion—the giant outdoor arena in The Woodlands that George named after his wife because of her love of the arts. The Houston Symphony, which George and Cynthia had long supported, provided the music, and a stream of colleagues, associates, and friends paid tribute from the heart of the city George had built.

Stephen Hawking couldn't attend, but he sent an audio tribute. He praised George's vision, wisdom, and persistence in establishing Texas A&M as a leading university in cosmology and physics research, and he reminisced about the time they spent together at Cook's Branch. "While none of us can match George's ingenuity in geophysical discovery, I am happy to say I still managed to beat him in wheelchair racing, even if it was only by a narrow margin," Hawking said. "Through sheer hard work and dedication, he leaves behind an extraordinary legacy. It can be said of very few people that they changed the world, but George Mitchell is among those few."[12]

George may have changed the world, but he didn't let the world or his success change him. "I am proud of you for not allowing your affluence to destroy your personality and the character you have always kept," a childhood friend, M. A. Caravageli, wrote in 1991. "Where does your true wealth lie? I would say it lies in the innate you, your God-given mind, your vision, your courage and your faith to accomplish the most that you can in your lifetime."[13]

CHAPTER 26

Benefits and Backlash

I stand just outside the fence that surrounds the C. W. Slay #1 and look toward the suburban homes, creeping from the horizon. As we get back into the truck, I point out the houses to my Barnett Shale guide, Jay Ewing, a former Mitchell Energy field manager who works for Devon. It's easy to see why there's been so much backlash over fracking, I say. Look at how close the houses are. For decades, most domestic drilling was done on unpopulated expanses in places like West Texas, not a stone's throw from people's back porches. Ewing nods. The buildings weren't there when the C. W. Slay was drilled in 1981, he points out. Fracking didn't encroach on the homes; the homes encroached on fracking. Then, as an afterthought, he says that neighborhoods like the one near the C. W. Slay—master-planned communities—are cropping up all over North Texas. I point out that The Woodlands became a model for such developments, and while most don't plan with the precision that George did, many of his ideas have been copied in real estate just as they were in oil and gas.

Both of George's biggest endeavors represent progress, and both are steeped in controversy. Fracking has reduced oil and natural gas prices, transforming the US energy landscape and the world of petro-politics. Yet it has come at a cost to the environment and communities. Sustainable development has created more livable neighborhoods and greater harmony with nature even as it has added to the problems of urban sprawl and increased demand for electricity and gasoline. Both of George's passions have had a profound effect on the lives of millions of Americans, feeding both the pull of and the resistance to progress.

◈

Five hours of meetings in Vienna on November 27, 2014, showed clearly how George had changed the world. The venerable dean of the global oil markets, Saudi Arabia's Ali bin Ibrahim al-Naimi, emerged from the closed-door session and declared that the Organization of the Petroleum Exporting Countries would maintain its oil production. Within minutes, the price of crude plunged. By the end of the year, it had tumbled more than 50 percent, slipping below fifty dollars a barrel for the first time in five years. In America, the price of gasoline fell in tandem, ultimately sending pump prices below two dollars a gallon. By early 2015, the Saudis' goal seemed clear: let prices collapse in hopes of breaking the oil boom in the United States. In a historic turnaround, the world's biggest oil consumer was producing so much of its own energy that it needed far fewer Saudi imports.

For four decades, American presidents had vowed to free the United States from OPEC's grip. They pledged to meet the country's energy needs without relying on hostile nations that could hold the United States hostage for oil. They spent billions developing alternatives—biofuels, solar power, wind power, hydrogen fuel cells, and even energy from sea algae. Yet for most of that time, American diplomats strengthened ties with the Saudi kingdom. When rising prices threatened to crush the American economy, presidents from Ronald Reagan to Barack Obama traveled to the kingdom with what amounted to a simple plea: please pump more. America had long ago ceded its dominance in oil production to the Middle East. US oil output peaked in 1972, and that was after more than a decade of the rising imports that George and Johnny had warned about during the 1960s. America's biggest oil companies, seeking ever-larger returns, had cast their fortunes abroad, developing reserves in other countries. Some of those countries would later use the reserves against America and other developed nations that had become dependent on an insatiable thirst for crude.

But as America recovered from the financial crisis of 2008 and 2009, the country's energy outlook changed. Thanks to fracking, America became a major oil producer, not just the biggest consumer. OPEC's loosening grip on the world oil market grew directly from Mitchell Energy's relentless science experiment on the plains of North Texas that had begun thirty-one years earlier. By the time OPEC members met in 2014, US oil producers were fracking for oil and gas all across the country, unlocking major new reserves in North Dakota and revitalizing fields in West Texas and Colorado. That year, American oil production grew more than in any of the previous one hundred years, an abundance that enabled the country to reduce imports significantly. In 2015, not only did imports decline, but the United States was bringing in five times as much oil from Canada as it was from Saudi Arabia.

Then, as quickly as fracking changed the global outlook for energy, it began to succumb to its own success. The surge in US production, combined with weakening demand from countries like China, pushed prices well below thirty dollars a barrel by 2016. The level was so low that most American producers could no longer afford to keep their rigs running. Across the energy industry, companies cut tens of thousands of jobs in 2015 and 2016. The record profits oil companies had reported just a few years earlier evaporated. In their wake came stunning losses. The industry found itself once again in the teeth of a devastating bust.

On a larger scale, however, lower energy prices boosted the overall economy. In 2016, cheaper gasoline alone saved Americans more than $115 billion, or about $550 per licensed driver, according to the American Automobile Association.[1] Meanwhile, the increased oil and gas production that fracking unleashed boosted the average American household's net worth by $1,900, according to a study by the University of Chicago's Energy Policy Institute and the Massachusetts Institute of Technology.[2] Some experts estimate that American consumers spent about $180 billion a year less on gasoline in 2016 than they did three years earlier.[3]

The changing production trends altered the geopolitical landscape. America's new oil and gas abundance increased its diplomatic muscle in the Middle East. President Obama and his team of diplomats hammered out a historic nuclear arms agreement with Iran, despite opposition from Saudi Arabia. (Obama's successor, Donald Trump, later rescinded the agreement.) The role that rising

US oil and gas production from fracking played in bringing Iran to the negotiating table was largely overlooked, but a decade earlier, a US president could not have afforded to anger the Saudis with such a deal, given America's dependence on the kingdom's oil supply. The United States shifted its domestic policies as well. In December 2015, President Obama authorized the first exports of domestic oil since 1975.

By late 2017, the economic upheaval that fracking had caused in Saudi Arabia had forced the kingdom to consider investing in West Texas shale production—the first time the Saudis had ever sought to buy drilling interests outside their own borders.[4] In 2018, the United States became the world's largest producer of both oil and natural gas. OPEC, long accustomed to setting the rules in the global oil market, was considering a lobbying campaign to influence US lawmakers. Increased production from fracking has formed the foundation of President Donald Trump's "energy dominance" policy.

Fracking, in turn, has helped the United States take the lead on other efforts to reduce climate change. The cheap and abundant natural gas supplies unleashed by fracking have created an economical alternative to coal for generating electricity. In 2015, representatives of 195 nations adopted a sweeping agreement that committed most countries on the planet to reducing the emission of greenhouse gases, primarily carbon. The next step in that effort comes from using more natural gas as a generation source or a backup for renewables such as wind and solar power. The abundance of natural gas, in other words, encourages a divorce between continued prosperity and fossil fuels.

This shift to natural gas has profoundly affected the electric utility industry. In 2007, a gaggle of Wall Street and private equity firms—Goldman Sachs, KKR, TPG Capital, Lehman Brothers—paid more than $8 billion for TXU Corp., the main utility for Dallas and North Texas. At the time, the state's electricity prices were tied to natural gas because of Texas's complex efforts to deregulate the utility business. The buyout's engineers believed that natural gas prices would keep rising, taking electricity rates with them. TXU generated much of its power from burning coal, which was still cheaper than gas at the time. The buyers hoped to make a tidy profit on the difference between the two fuel sources. But a year later, as fracking unleashed a flood of new natural gas, prices slid. By 2009, the company, which changed its name to Energy Future Holdings, couldn't repay debt from the buyout. In 2014, it filed one of the largest bankruptcies in US history.[5]

As it recovers from the bust, the US energy industry has refined fracking, building on George's work by making it more cost effective and reducing its environmental impact. Drilling is more automated, resembling a manufacturing process, with robotic rigs that "walk" from one well site to the next. With this new technology, companies can restart production in a matter of months, enabling producers to respond quickly to rising prices. These advances have reduced the need for some of the most expensive and environmentally threatening oil production, such as oil sands in northern Canada and offshore drilling in the Arctic.

Such benefits, though, have come at a cost. As the use of fracking has spread, residents in parts of North Texas—including the Dallas suburb of Irving—and Oklahoma have experienced a sharp rise in earthquakes. Before 2008, Oklahoma had one or two quakes of greater than 3.0 magnitude each year. In 2014, it recorded 585—three times the rate in California. William Ellsworth, a research geologist at the US Geological Survey, told the *New Yorker* in 2015: "We can say with virtual certainty that the increased seismicity in Oklahoma has to do with recent changes in the way that oil and gas are being produced." The quakes aren't coming from fracking itself, but from a by-product of the process. In addition to producing oil and gas, wells also produce water from the same formations, and that water rises to the surface along with the hydrocarbons. Once separated, this produced water is returned to the ground, usually by pumping it into disposal wells—depleted reservoirs that serve as huge underground storage tanks. (In addition, water injected from the surface as part of the fracking process may also be disposed of in this way, but the volume of injected water and fracking fluid represents about 10 percent of the volume of produced water, which can continue to rise to the surface for the entire life of the well.)[6] To sequester this water quickly, it's injected into the disposal well under tremendous pressure, and the volume of water, combined with the pressure of the injection, can lead to tremors.[7] As drilling increases, the need for disposal wells rises in tandem, causing more seismic activity and the potential for earthquakes in drilling areas.

There's another problem: shale formations produce oil and gas differently from conventional reservoirs. The density of the shale requires more wells, drilled closer together. For example, in a limestone reservoir on the Gulf Coast, a single well might draw oil or natural gas from several hundred acres. A shale well, however, might produce from as little as forty acres.[8] As a result, companies plan more wells per acre than with conventional drilling. Horizontal drilling techniques can reduce some of the impact of chockablock wells, but as George warned, more drilling means more environmental damage and more angry property owners.

Financing all those wells may become more difficult if interest rates rise. Energy insiders still debate the same issue that George and Bill Stevens did almost two decades ago: Is fracking economically viable? Without a steady influx of capital to finance new wells, fracking may not be sustainable, especially if prices fall globally. Some of the most active drillers in shale spend more than their cash flow on drilling costs.

Because many major shale plays, including the Barnett, are closer to homes than oil companies are accustomed to operating, few have bothered to take the care that George did when he hid his rig in a warehouse to drill Lafitte's Gold in Galveston. Widespread reports of water contamination in Dimock, Pennsylvania, and the fears stirred up by the *Gasland* film reverberated in communities across the country. Vermont banned fracking in 2012; New York followed two years later. Nine other states proposed bans but didn't enact them.[9] Ironically, one of the first such bans came less than thirty miles from where the fracking boom started—in Denton, Texas, which sits atop the Barnett Shale. At the time, the city had 272 active wells within its boundaries. Residents complained about rig noise, foul-smelling air, and traffic congestion from fleets of trucks rolling to and from well sites. Under Texas law, most suburban homeowners don't own the rights to the minerals under their land, which means that unlike the Galveston residents who participated in Lafitte's

Gold or the Wise County lease holders that George battled in the 1990s, the Denton homeowners received no economic benefit from drilling in their neighborhoods. Fifty-nine percent of voters in the North Texas college town supported the ban, after companies drilling in the area repeatedly ignored city regulations that required 1,200 feet between homes and well sites. (The companies argued that their permits, issued before the city ordinance, allowed them to drill closer, so they did.) After Denton banned fracking, Texas legislators overruled voters, saying that only the state can set limits on oil and gas operations. That sort of heavy-handed response has further tarnished fracking in the mind of the public.

Had George lived to see such controversy, he would have found it another challenge to overcome. Already, some oil companies are reducing the water they use and improving their methods for wastewater disposal. More of them are now disclosing the chemicals used in the fracking process after years of refusing to do so. Some are working with communities to reduce noise and traffic congestion.

The pros and cons of fracking complicate George's legacy. Did he inspire an economic boon or unleash an environmental scourge? As with all great inventions, the answer isn't one or the other, but both. George didn't invent a device as did Thomas Edison or Alexander Graham Bell. He perfected and adapted existing technology when no one else was close to such a breakthrough. He applied his entrepreneurial zeal and persisted for almost two decades, essentially betting his fortune on being right when all available science and knowledge said he was wrong. In that way he was as prescient and persistent as Edison or Bell. Fracking as we now know it changed all assumptions about the economics of extracting oil and gas from shale. "It took all those years of my dad not being constrained by the facts," Todd Mitchell said.

Cheaper, abundant energy can make countries worldwide richer, food safer and less expensive, industry more productive, and electricity and clean water available in corners of the earth where they previously didn't exist. But the spread of fossil fuels will increase carbon levels in the atmosphere over time, leaving the planet searching for ways to rein in the detrimental impact of such progress.

◈

The same duality hovers over George's legacy as a pioneer of sustainable development. The Woodlands may have been designed as a livable forest, at odds with most of the construction conventions around Houston, but George never intended for his city to remain separate from Houston forever. In fact, he asked Houston planners to include The Woodlands in the city's extraterritorial jurisdiction. "That means we don't run away from Houston's problems," he said in 1981. "In 10 or 15 years, when our tax base equals our bond retirement, Houston will annex The Woodlands, and it should."[10]

At the same time, he built The Woodlands on unincorporated land in Montgomery County, and he ran it through a private company that continues to oversee much of the development today. Despite his efforts to make The Woodlands a place to live and work, he knew its success depended on its proximity to Houston. And as such, it extended the sprawl of a metropolitan area that today encompasses more land area than the state of New Jersey. George had convinced the city of Houston and Harris County, which surrounds it, to support projects such as a major toll road from just south of The Woodlands into downtown. As a payoff for those investments, he offered The Woodlands' tax base to Houston.

The more The Woodlands grew, however, the more residents opposed the idea of being absorbed into Houston. They had no interest in helping the bigger city solve its problems. The private-company control had always resulted in a sort of benevolent paternalism, but residents drew the line at George's holistic view of addressing urban ills. In 2007, The Woodlands Development Corp. struck a deal to forestall annexation for fifty years. Now partially governed by an elected township council, The Woodlands is studying what it must do to incorporate as a stand-alone city. Of all the changes since George sold it, nothing violated his vision for the community's future more than turning away from Houston, a city to which George felt a great debt for his success. "The Woodlands was never meant to be a separate community because that type of development pattern is what is destroying our cities around the country," he said. "Separate little communities that parasite off a major city are destructive. The talented people and the tax base don't contribute toward the city. The people don't make an effort to make the city work."[11]

These days, The Woodlands is more focused on its own issues, such as traffic congestion and self-governance. In 2016 voters threw out old-guard council members—many of whom had ties to George or claimed to support his plan for the community—in favor of leaders who advocated greater local control. During the bitter election season, both sides invoked George's vision for his so-called eleventh child. But The Woodlands had years ago moved past George's ideas.

As of 2016, it had 112,000 residents—more the double the population when he sold in 1997.[12] It has changed owners three times since then, and the clash between George's vision and that of its later developers is obvious. Four days after he announced the sale, George sent a letter to residents assuring them that little would change. "I think it will be hard to tell that anything is different," he wrote.[13] But as the development has matured, the differences have become more obvious. The early villages that George built—Grogan's Mill, Cochran's Crossing, Panther Creek—adhere to the ideals of wide greenbelts, tree-filled lots, and ample space between houses. In the newest neighborhoods, trees have been clear-cut and houses are stamped on tiny lots, creating rows of cookie-cutter neighborhoods that look like any other American suburb.

What's more, the lots in Creekside, the newest village, were filled with dirt, raising them to remove them from the federal flood plain maps. While the process was legal, it left the area vulnerable to flooding following the massive rainfall from Hurricane Harvey in August 2017.[14] (The rest of The Woodlands had few flooding issues, despite being inundated with more than fifty inches of rain.)

Demographically, The Woodlands has fallen short of George's altruistic vision of a city inclusive of all races, religions, and social and economic classes. Eighty-eight percent of residents are Caucasian, and 68 percent are married.[15] Its jobs are primarily white collar. It has the same exclusivity that comes with any place that people find desirable to live: the more people who want to live there, the more the cost of housing rises, making it difficult for lower-income home buyers to move in. By early 2017, the median home price in The Woodlands was $428,750, compared with $175,200 for the greater Houston area.[16] It added the gated enclaves that George loathed, creating exclusive homes for rap stars, pro sports figures, and wealthy Mexicans fleeing violence in their native country. Of the more than 37,000 housing units that The Woodlands reported in the 2010 Census, only about 1,000 are subsidized for low-income residents.[17] George had hoped to keep the level of subsidized housing high enough that the community would become more diverse economically and ethnically, not less. He predicted that attracting minorities at a level that would match the overall demographics of Houston would take forty years. Instead, the demographics of The Woodlands fell behind the shift that occurred in Houston, which evolved into the most ethnically diverse major city in America. The Woodlands' median family income soared to $98,675, almost double the national average.[18] In other words, the community has become the same sort of exclusive retreat that whites sought in the 1960s as they fled the decaying inner cities—albeit better planned and with more appealing amenities.

While envisioned as a community for all—"a home for the janitor and a home for the president"—The Woodlands also shows the difficulty of marrying economics and social consciousness. "I'm not sure a janitor can live in The Woodlands now," said David Hendricks, who worked on the project in the early 1970s. "If you live in The Woodlands, you pay a premium for being in The Woodlands."

Designed as a remedy to urban sprawl, The Woodlands now accelerates it. People who work there commute from communities farther north, extending the concrete sea of Houston past once-distant cities like Conroe. Yet in this, too, George was a pioneer. In the more than forty years since the development of The Woodlands, as much as 85 percent of the residential and job growth in the Houston area has fallen outside city limits.[19]

While George didn't solve America's urban ills, he did expand the concept of livability. In 2015, The Woodlands ranked as the most livable city in Texas, based on its low rates of crime, unemployment, and commute time and its high walkability, educational levels, and homeownership.[20]

"I don't know what the answer to the problem of the cities is, but we have to keep looking for it," George once said. "We have to try alternatives. The Woodlands is one of them. It's not a subdivision, it's a self-supporting unincorporated area that will eventually become a contributing part of Houston's tax base. There *are* solutions to our problems, and we can find them. We have to. If we don't try, then we'll have failed our children and the future."[21]

◈

The worlds of energy production and sustainable development may seem at odds in today's debates about climate change, but to George, they went hand in hand. He never saw them in conflict, but as ideas worth pursuing on their individual merits. A cheaper, cleaner-burning, and more abundant energy resource such as natural gas offers the opportunity for continued growth at lower cost to the environment. It also provides a foundation for fuels with less environmental impact, such as wind and solar power. Planned development was the bedrock of his quest to mitigate the social ills brought about by race, poverty, and pollution. More important, though, George's achievements in energy and sustainable development, like his support for physics and other interests, reflect a lifelong intellectual curiosity that couldn't be limited to one set of problems.

His life is a reminder that success in business isn't defined by just stock price or profit. By those measures, he was a great oil finder who turned out to be a successful businessman. He built a company and sold it at just the right time, making billions in the process. The first thirteen wells he drilled in the Frustration Fields are still paying royalties to George's children. While that sort of financial and geological success would be enough for most oilmen, George believed business should aspire to something grander—to making society better. To him, business success wasn't the goal, but the beginning from which far greater endeavors could be launched. Most other companies wouldn't, and didn't, pursue projects like fracking. Nor would they have done The Woodlands the way he did it, sacrificing years of profitability for livability. "He wasn't a crazy mad scientist, but he could actually visualize the science to see what it could do," said Joe Greenberg, the Alta Resources CEO who worked with George in the final years of his life.

George thought The Woodlands would be his legacy. Others now say it is fracking, and some believe it is both. Either would be significant for the son of an illiterate goatherd.

His legacy transcends his achievements. It is rooted in a more important principle: a love of ideas. For George, making money and creating jobs were important, but only if the business offered something more—only if it drew out the best in people and gave them the freedom to realize their individual potential. "He fostered an atmosphere that let people think; that let good ideas come out, and he was willing to try them," his former spokesperson Tony Lentini said. "George was an example of what people can do if they are good of heart and aren't afraid."

His willingness to tackle big, seemingly insurmountable problems was infectious. Because he believed something could be done, whether it was fracking or sustainable development or drilling under the city streets of Galveston, his determination drew others to his vision, geologist Howard Kiatta said. "He inspired you to do the impossible."[22]

Acknowledgments

I am grateful to the Mitchell family for their cooperation in the writing of this book. From the beginning, they understood the importance of preserving the integrity of the project, and while they all contributed hours of their time, they agreed that they would have no editorial control. Sheridan Lorenz and Meredith Dreiss were particularly helpful in coordinating access to George Mitchell's papers, offering family correspondence and photos, and encouraging their siblings to participate in the process. Todd and Wendy Mitchell allowed me to spend the day with them in Aspen and graciously took me to dinner at Wendy's wonderful restaurant, Meat and Cheese. Scott Mitchell agreed to an impromptu interview that went on for hours. All ten of George and Cynthia's children allowed me to intrude on their annual Thanksgiving gathering at Cook's Branch in 2015.

Sheridan, Todd, Dan Steward, Ed Fry, and Joe Greenberg all reviewed, at my request, all or portions of the manuscript and offered suggestions and corrections.

George Mitchell didn't want a biography written and instead commissioned Mitchell Energy's public relations manager, Joseph Kutchin, to write *How Mitchell Energy & Development Corp. Got Its Start and How It Grew* (Universal Publishers, 2001). The book is invaluable for its lengthy interviews with company employees, many of whom are no longer living.

Courtney Herrington, Nancy Stucky, and the Mitchell Family Corporation ensured I had access to an office and a copier and worked with the Mitchell family to gather George's papers in a single place. Brett Holmes with the Mitchell Family Foundation was a critical first point of contact in getting this project started and supporting it throughout the process.

Denise Manning, the library operations coordinator at Tarpon Springs Public Library, went above and beyond, not only searching the library's genealogical resources but also contacting the local

Greek Orthodox church for any leads on Katina Eleftheriou's early years in America. Gayle Davis-Cooley shared her reflections on the Mitchells' years at Trinity Episcopal Church in Houston, as well as information mined from the church archives.

John Poretto helped line up key interviews with Devon Energy executives, including former chair Larry Nichols, and he arranged my tour of the Barnett Shale with Jay Ewing. Yes, Jay's middle initial is "R.," so I was shown one of the most important sites in the history of the energy industry by none other than J. R. Ewing. Jay, however, was far more gracious than his TV namesake, and I appreciate both his time and his patience.

John Williams came through with an unexpected interview with James A. Baker III, who was helpful in nailing down a few points about the history of tennis in Houston. The ever-reliable Jay Rosser set up a lunch with industry legend Boone Pickens, who shared his reflections on the natural gas business, even though he didn't know George Mitchell well. Cary Lindsay arranged coffee with Coulson Tough, whose recollections were both valuable and enjoyable.

I also must thank my fellow journalists and former colleagues. Frank Smith dug through the musty microfilm archives in the bowels of the *Dallas Morning News* to track down a decades-old story on the Frustration Fields, and Mary Schlangenstein saved me countless hours of tedium by providing Bloomberg data on Mitchell Energy's daily stock prices.

Gail Roche, my longtime editor at *Bloomberg News*, gave the manuscript a good scouring that, when I received it, prompted only a couple of walks around the yard. As always, her insights made the manuscript sing and saved me from my own bad habits. David Kaplan read the manuscript with the perspective of a native Houstonian and offered valuable insights.

Jim Blackburn, Art Berman, and Lillian Espinoza-Gala also provided important feedback.

I also must thank my colleagues at 30 Point Strategies, particularly principals Adam Levy and Noam Neusner, for their flexibility and support. Without their understanding, none of this would have been possible.

Jay Dew, Gayla Christiansen, Christine Brown, Katelyn Knight, and the rest of the team at Texas A&M University Press once again have been a pleasure to work with, and I am glad that this project found the perfect home. Laurel Anderton gave the manuscript a final polish.

And last but certainly not least, Laura, my wife of more than thirty years, deserves the most credit of all. Once again, she found herself a "book widow," yet she showed nothing but support and understanding. She also provided the first read of the manuscript and offered invaluable perspective and advice as a general reader.

Notes

Preface

1. Alando Jones Ballantyne to George P. Mitchell (GPM), May 21, 1984.

2. Robert Levy, "Wildcatters Hunt Texas Oil," *Boston Globe*, undated clipping among GPM's private papers.

3. Kim Hogstrom, "George Mitchell: Galveston, Gambling and the Persian Gulf," *DBA Magazine*, March 1991, 27.

4. Dancie Ware, interview by the author, April 7, 2016.

5. Mary Lance, "The Woodlands: Is It Working?" *Southwest Airlines Magazine*, August 1982, 118.

Chapter 1

1. George Mitchell to Cynthia Mitchell, undated letter.

2. Lisa Gray, "Bayous for Dummies," *Houston Chronicle*, October 17, 2010, accessed January 17, 2017, http://www.chron.com/life/article/Bayous-for-Dummies-1706294.php.

3. "Allen's Landing, TX," *Handbook of Texas Online*, Texas State Historical Association, https://www.tshaonline.org/handbook/online/articles/hvabg.

4. Walter McQuade, "Good Living in Houston: At Home beside the Bayou," *Fortune*, July 1, 1966, 111.

5. "US Dry Natural Gas Production," US Energy Information Administration, https://www.eia.gov/dnav/ng/hist/n9070us2a.htm.

6. "Natural Gas Expected to Surpass Coal in Mix of Fuel Used for US Power Generation in 2016," US Energy Information Administration, https://www.eia.gov/todayinenergy/detail.php?id=25392.

7. "Cushing, OK WTI Spot Price FOB," US Energy Information Administration, https://www.eia.gov/dnav/pet/hist/LeafHandler.ashx?n=pet&s=rwtc&f=m.

8. "Fueling the Future with Natural Gas: Bringing It Home," IHS CERA, January 2014, https://www.fuelingthefuture.org/assets/content/AGF-Fueling-the-Future-Study.pdf.

9. Anthony Watts, "USA Meets Kyoto Protocol Goal—Without Ever Embracing It," *Watts Up with That?* (blog), April 5, 2013, https://wattsupwiththat.com/2013/04/05/usa-meets-kyoto-protocol-without-ever-embracing-it/; "Bush:

Kyoto Treaty Would Have Hurt Economy," Associated Press via NBCnews.com, June 30, 2005, http://www.nbcnews.com/id/8422343/ns/politics/t/bush-kyoto-treaty-would-have-hurt-economy/#.W01aSbi1vRZ.

10. Anna Hirtenstein, "Record Green Power Installations Beat Fossil Fuel for First Time," *Bloomberg*, October 25, 2016, https://www.bloomberg.com/news/articles/2016–10–25/record-green-power-installations-beat-fossil-fuel-for-first-time.

11. Ralph Bivins and Andrea Ahles, "Sale Makes Some Residents Uneasy," *Houston Chronicle*, June 14, 1997, 3C.

12. Joseph W. Kutchin, *How Mitchell Energy & Development Corp. Got Its Start and How It Grew*, updated ed. (Boca Raton, FL: Universal Publishers, 2001), 376.

13. Michael Davis, "Questions Raised about Mitchell's Next Steps," *Houston Chronicle*, June 14, 1997.

14. Anne Pearson to GPM and Budd Clark, memorandum, July 30, 1997.

15. "Population of The Woodlands, TX," TheWoodlandsTX.com, https://www.thewoodlandstx.com/demographics/.

Chapter 2

1. "George Phydias Mitchell, Volume I, The Early Years," family photo album, undated.

2. "Introducing Peloponnese," Lonely Planet, http://www.lonelyplanet.com/greece/the-peloponnese/introduction.

3. Mitchell family photo album.

4. Socrates Petmezas, "History of Modern Greece," in *About Greece*, edited by G. Metaxas, 7–41 (Athens, 2002).

5. *The Greek Texans* (San Antonio: University of Texas at San Antonio Institute of Texan Cultures, 1974), 1–2.

6. GPM, interview by Paul Wiesenthal, unpublished transcript, June 16, 1988, 35.

7. Kutchin, *How Mitchell Energy*, 364.

8. Kutchin, 443.

9. *Greek Texans*, 7.

10. Kitty Kendall, "'Papa Mike' Was Laborer on California Railroad," *Galveston News Tribune*, undated clipping from Mitchell family papers. Family members tell several variations of this story. In some, Savvas asks the paymaster his name, and the paymaster replies, "Mike Mitchell." Savvas then says, "That's my name too." In others, the paymaster assigns Savvas the name Mike Mitchell. George told both versions of the story at various times. I have chosen to use Mike's own recounting.

11. Kutchin, *How Mitchell Energy*, 364.

12. Mitchell family photo album.

13. Kutchin, *How Mitchell Energy*, 364; *Greek Texans*, 1–2.

14. "The Courts" (court dockets), *Galveston Daily News*, October 4, 1905, 7, Newspapers.com.

15. "Mitchell and Chiacos Brothers Were Released under $2,000 Bonds, *Houston Post*, September 1, 1914, 9, Newspapers.com.

16. Advertisement for People's Restaurant, *Houston Post*, May 8, 1908, 14, Newspapers.com.

17. Kendall, "'Papa Mike.'"

18. GPM, interview by Pamela Mitchell McGuire, 1978.

19. Christie Mitchell, "To the End, Mike Had a Will to Live," *Galveston Daily News*, December 27, 1970, 10, Newspapers.com.

20. Kendall, "'Papa Mike.'"

21. Mitchell family photo album.

22. The story is unlikely because Mike was barely able to read, even in Greek. It seems unlikely that Katina's arrival in Florida warranted a magazine story in Texas, and no one in the family can recall the name of the publication.

23. Christie Mitchell, "To the End," 10.

24. Passenger list, *S.S. Patris*, departed Piraeus on May 22, 1910.

25. Kirk Mitchell and several other family members suggested this was the most likely story of how Mike and Katina met during interviews by the author on November 27, 2015.

26. Harris County marriage license records (as "M. Saocks Paraskevopulos" and "Katina Elefteins"), accessed December 28, 2016, https://www.ancestry.com/interactive/9168/45607_b299041-01858?pid=92633; "Mike Mitchell, for Whom Campus Was Named, Dies," *Galveston Daily News*, December 26, 1970, 1, Newspapers.com.

27. Christie Mitchell to Maria Mitchell, February 20, 1940.

28. Dio Adallis, *Adallis' Greek Merchants' Guide*, 1913, 28 (from Mitchell family archives).

29. Kendall, "'Papa Mike.'"

30. Adallis, *Adallis' Greek Merchant Guide*, 28.

31. Joanne Harrison, "The Making of a Billionaire," *Houston City Magazine*, March 1981, 52.

32. Erik Larson, *Isaac's Storm: A Man, a Time, and the Deadliest Hurricane in History* (New York: Vintage Books, 1999), 265.

33. *Greek Texans*, 4.

34. Kutchin, *How Mitchell Energy*, 371.

35. Christie Mitchell to Maria Mitchell Ballantyne, undated (about 1962).

36. "Widow of Victim of Main Street Fire Gets Check from Magnolia Park Land Co., in Settlement of Death Claim," *Houston Post*, November 1, 1914, 25, Newspapers.com.

37. "Two Men Who Lost Their Lives When Burning Building Was Wrecked by an Explosion Have Been Identified as Joe Nader and Maynard Upton," *Houston Post*, August 24, 1914, 1, Newspapers.com.

38. "Two Men Who Lost Their Lives," 1.

39. "Mitchell and Chiacos Brothers Released," *Houston Press*, September 1, 1914, 9, Newspapers.com.

40. "Mitchell and Chiacos Brothers," 9.

41. "Three Men Give Themselves Up," *Galveston Daily News*, November 3, 1914, 3, Newspapers.com.

42. "Several Indicted Men Gave Bonds on Monday," *Houston Post*, November 3, 1914, 9, Newspapers.com.

43. "Criminal District Court" (docket report), *Houston Post*, December 29, 1914, 16, Newspapers.com.

44. Christie Mitchell, "Mitchell Family to Get Together," *Galveston Daily News*, March 27, 1977, 61, Newspapers.com.

45. Kirk Mitchell, interview by the author, November 27, 2015.

46. "Cheaper Rents on Main Street," *Houston Post*, March 14, 1915, 24, Newspapers.com.

Chapter 3

1. Kendall, "'Papa Mike.'"

2. "Births," *Galveston Daily News*, April 7, 1912, 12. Although his birth certificate says "Christ," that is likely the anglicized version of the Greek name "Christos," which was the name of Mike's father. Greek tradition called for naming the first son after his paternal grandfather.

3. "US Social Security Death Index 1935–2014," Ancestry.com, accessed December 28, 2016. In keeping with Greek tradition, the second son was named after the maternal grandfather. Katina's father was Ioannis Eleftheriou.

4. Mitchell family photo album.

5. Harrison, "Making of a Billionaire," 53.

6. Kirk Mitchell, interview.

7. Christie Mitchell, "Mitchell Family," 61.

8. George Mitchell, conversation with Pamela Mitchell McGuire, undated. In Greek culture, the first and second sons are traditionally named after their grandfathers. The third son is named by the godfather.

9. Anne Belli Gesalman, "George Mitchell, a Work Ethic Inspired by the Past Extends to the Future," *Dallas Morning News*, February 9, 1997, 4E.

10. 1910 US Census data show that Mike lived at 2214 Ave. D, Ward 4, in the Strand. The 1920 census shows 2224 2nd Ave., Ward 9.

11. Gary Cartwright, *Galveston: A History of the Island* (Fort Worth: Texas Christian University Press, 1998), 207.

12. Harrison, "Making of a Billionaire," 53.

13. Pamela Mitchell McGuire, interview by the author, November 27, 2015; Kutchin, *How Mitchell Energy*, 366.

14. Kutchin, 365.

15. Kirk Mitchell, interview.

16. GPM to Jimmy Lampis, January 22, 1962.

17. Kirk Mitchell, interview.

18. Venita Van Caspel, interview by the author, September 24, 1985, 2.

19. Scott Mitchell, interview by the author, March 14, 2016.

20. Cartwright, *Galveston*, 208.

21. Cartwright, 209.

22. "Barker Enjoins Places on Beach," *Galveston Daily News*, July 31, 1929, Newspapers.com.

23. James Presley, *A Saga of Wealth: The Rise of the Texas Oilmen* (Austin: Texas Monthly Press, 1983), 243.

24. Cartwright, *Galveston*, 212.

25. Nicole T. Boatman, Scott H. Belshaw, and Richard B. McCaslin, *Galveston's Maceo Family Empire: Bootlegging & the Balinese Ballroom* (Columbia, SC: History Press, 2014), 57–69.

26. Robert Draper, "Big Fish," *Texas Monthly*, May 1997, accessed August 10, 2015, http://www.texasmonthly.com/articles/big-fish/.

27. GPM to Maria Mitchell, October 12, 1939.

28. Sheridan Lorenz, interview by the author, October 28, 2015.

29. Lorenz, interview, October 28, 2015.

30. McGuire, interview.

31. Mitchell family photo album.

32. *Galveston Daily News*, undated clip from about 1935.

33. Kutchin, *How Mitchell Energy*, 366.

34. William L. Payne, William L. Payne II, and Mary Elizabeth Payne, *44 Different and Proven Ways for Kids to Make Money* (1999), 35 (from Mitchell family archives).

35. Ware, interview.

36. Mitchell family photo album.

37. "George Mitchell: The Energy Man," *Greek Accent*, September/October 1986, 14.

38. "Trial of Large Damage Claim Opens in Court," *Galveston Daily News*, April 19, 1949, 1, Newspapers.com.

39. Kutchin, *How Mitchell Energy*, 365.

40. Harrison, "Making of a Billionaire," 53–54.

41. Katina Mitchell death certificate, Texas Bureau of Vital Statistics.

42. Gesalman, "George Mitchell," 4E.

43. Mitchell family photo album.

44. GPM to Maria Mitchell, October 12, 1939.

45. Lorenz, interview, October 28, 2015.

46. Kutchin, *How Mitchell Energy*, 365.

47. Gesalman, "George Mitchell," 4E.

48. Harrison, "Making of a Billionaire," 53.

49. Kutchin, *How Mitchell Energy*, 368.

50. Harrison, "Making of a Billionaire," 54.

51. Kendall, "'Papa Mike.'"

52. Christie Mitchell, "To the End."

53. Lillian E. Herz, "Christie Mitchell Spreads Word of the Isle's Charms," *Galveston News-Tribune*, August 23, 1964, 3, Newspapers.com.

54. Presley, *Saga of Wealth*, 241–42.

55. Kutchin, *How Mitchell Energy*, 367.

56. Kutchin, 368.

57. Presley, *Saga of Wealth*, 242.

58. Johnny Mitchell to Maria Mitchell, December 15, 1939.

59. Harrison, "Making of a Billionaire," 54.

60. Kutchin, *How Mitchell Energy*, 368.

Chapter 4

1. Daniel Yergin, *The Prize: The Epic Quest for Oil, Money & Power* (New York: Simon & Schuster, 1991), 244–46.

2. Lawrence Goodwin, *Texas Oil, American Dreams* (Austin: Texas State Historical Association, 1996), 2–3.

3. Kutchin, *How Mitchell Energy*, 369.

4. Harrison, "Making of a Billionaire," 55.

5. Kutchin, *How Mitchell Energy*, 369.

6. Kutchin, 370.

7. Kutchin, 371.

8. Boatman, Belshaw, and McCaslin, *Galveston's Maceo Family Empire*, 74.

9. Harrison, "Making of a Billionaire," 55.

10. Harrison, 55.

11. Harrison, 55.

12. Gesalman, "George Mitchell," 4E.

13. Harrison, "Making of a Billionaire," 55.

14. Kutchin, *How Mitchell Energy*, 372.

15. Kutchin, 371.

16. Christie Mitchell to Maria Mitchell, February 20, 1940.

17. Kutchin, *How Mitchell Energy*, 372.

18. "John 'Jarrin' John' Kimbrough," National Football Foundation, College Football Hall of Fame, accessed August 8, 2015, http://www.footballfoundation.org/Programs/CollegeFootballHallofFame/SearchDetail.aspx?id=30081.

19. Kutchin, *How Mitchell Energy*, 370.

20. G. B. Winstead, "Isle Netter at A and M Disproves Theory That Athletes Are Dumb," *Galveston Daily News*, date unknown (clipping from GPM's personal files).

21. "George P. Mitchell: Sixty Years of Tradition Continued," Texas A&M University press release, May 9, 2014, accessed November 12, 2015, http://engineering.tamu.edu/news/2014/05/09/george-p-mitchell-sixty-years-of-tradition-continued.

22. Mike and Christie Mitchell to GPM, telegram, undated.

23. Winstead, "Isle Netter."

24. Kutchin, *How Mitchell Energy*, 373.

25. Presley, *Saga of Wealth*, 242; Johnny Mitchell, *The Secret War of Captain Johnny Mitchell* (Houston: Gulf Publishing, 1976), 1–2.

26. GPM, US Army records.

Chapter 5

1. Sandy Sheehy, *Texas Big Rich: Exploits, Eccentricities, and Fabulous Fortunes Won and Lost* (New York: William Morrow, 1990), 360.

2. "Hex Rally," Texas Exes, accessed May 16, 2017, https://www.texasexes.org/about-texas-exes/history-and-traditions/ut-traditions/hex-rally; "A Look Back: Memorable Games and Historical Moments That Shaped the Texas A&M-University of Texas Rivalry," *Texas A&M Today*, accessed May 16, 2017, http://today.tamu.edu/2011/11/21/a-look-back-memorable-games-and-historical-moments-that-shaped-the-texas-am-university-of-texas-rivalry/.

3. "Hex Rally"; "A Look Back."

4. Gesalman, "George Mitchell," 4E.

5. Gesalman, 4E.

6. "Cynthia Woods Mitchell," paid obituary, *Houston Chronicle*, December 30, 2009, accessed October 18, 2016, http://www.legacy.com/obituaries/houstonchronicle/obituary.aspx?pid=137965860.

7. Gesalman, "George Mitchell," 4E.

8. US Census, marriage and birth records.

9. Lorenz, interview, October 28, 2015.

10. Lorenz, interview, October 28, 2015; US Census records.

11. City directory, Jacksonville, Illinois, which lists Loretta as "widowed."

12. *Crimson J*, yearbook for Newton Bateman Memorial High School, Jacksonville, Illinois, 1939.

13. "Underground Railroad," Jacksonville Area Convention & Visitors Bureau, 2014, accessed January 4, 2017, http://jacksonvilleil.org/wp-content/uploads/2014/08/UGRR-Brochure2014.pdf.

14. "Newton Bateman Memorial High School," Clio, accessed January 4, 2017, https://theclio.com/web/entry?id=28258.

15. "Cynthia Woods Mitchell," obituary.

16. *Crimson J* yearbook, personal autographed copy among Cynthia Woods Mitchell's personal papers. The autographs include numerous references to her moving to New York.

17. Lorenz, interview, October 28, 2015. US Census shows Loretta, Cynthia, and Pamela living in Houston by 1940.

18. *Who's Who among Students in American Universities and Colleges* (Tuscaloosa, AL: Randall-Reilly, 1943).

19. *Houstonian*, University of Houston yearbook, 1942.

20. Claudia Feldman, "Low-Key Billionaire," *Houston Chronicle*, *Texas* magazine supplement, June 22, 2003, 11.

21. Johnny Mitchell, *Secret War*, 9.

22. "Dickson Gun Plant," Texas State Historical Association, accessed November 12, 2016, https://tshaonline.org/handbook/online/articles/dmd01.

23. GPM to Jimmy Lampis, November 29, 1942.

24. Kirk Mitchell, interview.

25. GPM, interview by Paul Wiesenthal, 3–4.

26. Kutchin, *How Mitchell Energy*, 375.

27. Wedding announcement, *Houston Post*, November 21, 1943 (clipping in Mitchell family photo album).

28. GPM, undated retrospective for company newsletter.

29. GPM, US Army records.

30. Presley, *Saga of Wealth*, 242.

31. Genealogy notes from Pamela Mitchell McGuire, undated.

32. Kutchin, *How Mitchell Energy*, 375.

33. Kutchin, 376.

34. Harrison, "Making of a Billionaire," 69.

Chapter 6

1. Harrison, "Making of a Billionaire," 69.

2. Kutchin, *How Mitchell Energy*, 310–11.

3. "He's Half Wildcatter, Half Big Businessman," *Businessweek*, November 24, 1964, 3 (reprint from GPM's private papers).

4. Sheridan Lorenz to Alleyne Mitchell (Johnny's wife), September 10, 1996.

5. Presley, *Saga of Wealth*, 244.

6. Bryan Burrough, *Texas Big Rich* (New York: Penguin Books, 2009), 167–89.

7. Presley, *Saga of Wealth*, 244–45.

8. Kutchin, *How Mitchell Energy*, 142.

9. Kutchin, 377–78.

10. Harrison, "Making of a Billionaire," 69.

11. Lorenz, interview, October 28, 2015.

12. Clinton Woods, US Social Security death index, 1935–2014, accessed January 4, 2017, https://search.ancestry.com/search/db.aspx?dbid=3693.

13. The account of Clinton's attempts to reconnect with the girls is based on interviews with members of the Mitchell family, who in the late 1970s found an associate of Clinton Woods in Maryland who was familiar with his letter writing.

14. Van Caspel, interview, 3–4.

15. GPM, interview by Paul Wiesenthal.

16. "He's Half Wildcatter," 3.

17. Goodwin, *Texas Oil*, 217.

18. GPM, interview by Paul Wiesenthal; Lorenz, interview, October 28, 2015.

19. Presley, *Saga of Wealth*, 209–11.

20. "Robert Everett Smith," *Handbook of Texas Online*, Texas State Historical Association, accessed January 3, 2016, https://tshaonline.org/handbook/online/articles/fsm57.

21. Kutchin, *How Mitchell Energy*, 144.

Chapter 7

1. Kutchin, *How Mitchell Energy*, 17, 381.

2. GPM, interview by Paul Wiesenthal.

3. Kutchin, *How Mitchell Energy*, 381.

4. Dan Steward, *The Barnett Shale Play: Phoenix of the Fort Worth Basin; A History* (Fort Worth: North Texas Geological Society, 2007), 12–26.

5. Kutchin, *How Mitchell Energy*, 146.

6. John A. Jackson, *Wise County: 25 Years of Progress* (Dallas: Williamson Printing, 1972), 26–28.

7. Jackson, 32.

8. Dudley Lynch, "The Frustration Fields Gang Rides Again," *Dallas Morning News, Southwest Scene* magazine, December 10, 1972.

9. Lynch, "Frustration Fields."

10. Kutchin, *How Mitchell Energy*, 381.

11. Boatman, Belshaw, and McCaslin, *Galveston's Maceo Family Empire*, 100.

12. Jackson, *Wise County*, 42.

13. Lynch, "Frustration Fields."

14. Kutchin, *How Mitchell Energy*, 382–83.

15. GPM, interview by Paul Wiesenthal.

16. Lynch, "Frustration Fields."

17. Kutchin, *How Mitchell Energy*, 147.

18. Kutchin, 142.

19. Jackson, *Wise County*, 46.

20. Dan Steward, "George P. Mitchell and the Barnett Shale," *Journal of Petroleum Technology*, November 2013, 59.

21. Kutchin, *How Mitchell Energy*, 143.

22. Kutchin, 334.

23. Kutchin, 313.

24. The cost may have been higher. John Jackson says it cost $32 million. Budd Clark later told Joe Kutchin it was $35 million (Kutchin, *How Mitchell Energy*, 143).

25. Jackson, *Wise County*, 47–48.

26. Jackson, 48.

27. Kutchin, *How Mitchell Energy*, 140.

28. Kutchin, 283.

Chapter 8

1. Kutchin, *How Mitchell Energy*, 278.

2. Lorenz, interview, October 28, 2015.

3. Todd Mitchell, interview by the author, April 25, 2016.

4. Mark Mitchell, interview by the author, November 27, 2015.

5. Grant Mitchell, interview by the author, November 27, 2015.

6. Kirk Mitchell and Grant Mitchell, interviews.

7. Lorenz, interview, October 28, 2015

8. Grant Mitchell, interview.

9. Gayle Davis-Cooley, "A Tribute to Cynthia and George Mitchell," *The Window of Trinity Church* (parish newsletter), September 2013.

10. Ware, interview.

11. "George P. Mitchell: Sixty Years of Tradition Continued."

12. David Dillon, *The Architecture of O'Neil Ford: Celebrating Place* (Austin: University of Texas Press, 1999), 1–4.

13. Christie Mitchell, with Bill Cherry, *The Night Owl* (Galveston, TX: self-pub., Bill Cherry and GPM, 2011), 234.

14. Christie Mitchell, *Night Owl*, 6–11.

15. Christie Mitchell, "To the End," 10.

16. "Snug Harbor Fire Suit Seeking $130,000 Damages," *Galveston Daily News*, December 5, 1947, 20, Newspapers.com.

17. Christie Mitchell, "To the End," 10.

18. Todd Mitchell, interview, April 25, 2016.

19. Presley, *Saga of Wealth*, 245.

20. Van Caspel, interview, 5.

21. "Mitchell Energy and Development Corporation History," Funding Universe, accessed January 4, 2017, http://www.fundinguniverse.com/company-histories/mitchell-energy-and-development-corporation-history/.

22. Kutchin, *How Mitchell Energy*, 152.

23. Kutchin, 139.

24. "Bernard Francis (Budd) Clark," paid obituary, *Houston Chronicle*, December 18, 2005, accessed March 25, 2016, https://www.legacy.com/obituaries/houstonchronicle/obituary.aspx?n=bernard-francis-clark-budd&pid=15995751.

25. Kutchin, *How Mitchell Energy*, 138–40.

26. Kutchin, 145.

27. Kutchin, 140–41.

28. Kutchin, 141–42.

29. Kutchin, 141.

30. Kutchin, 148.

31. Harrison, "Making of a Billionaire," 69–70.

Chapter 9

1. Kutchin, *How Mitchell Energy*, 196.

2. Tom Purcell, interview by the author, March 16, 2016.

3. "Shuffling of Upstairs Personnel by C.M.&M. Co.," *Wise County Messenger*, May 28, 1959, 1, Newspapers.com.

4. Maggie Galehouse, "Last Hurrah for the Houston Club's Rusk Location," *Houston Chronicle*, December 30, 2012, accessed January 10, 2017, http://www.houstonchronicle.com/news/houston-texas/houston/article/Last-hurrah-for-the-Houston-Club-s-Rusk-location-4156030.php.

5. Bryan Smothers, interview by the author, January 25, 2017.

6. Kutchin, *How Mitchell Energy*, 202.

7. "Big Oil, Gas Deal Reported," *Corsicana Daily Sun*, January 1, 1963, 8, Newspapers.com.

8. "He's Half Wildcatter," 1–3.

9. Harrison, "Making of a Billionaire," 70.

10. Harrison, 70.

11. Kutchin, *How Mitchell Energy*, 308.

12. Kirk Mitchell, interview.

13. Kutchin, *How Mitchell Energy*, 288.

14. "Jade Oil Joining Houston Oilmen," *Odessa American*, January 7, 1960, 2.

15. "He's Half Wildcatter," 2.

16. "He's Half Wildcatter," 1.

17. Matthew Simmons, *Twilight in the Desert* (New York: John Wiley & Sons, 2005), 10–15.

18. Goodwin, *Texas Oil, American Dreams,* 81–97.

19. "He's Half Wildcatter," 1.

20. Johnny Mitchell, excerpt of letter to Jack Rathbone, chair, Standard Oil of New Jersey, in *Energy—And How We Lost It* (Houston: Gulf Coast Publishing, 1979), 21–22.

21. "He's Half Wildcatter," 3.

22. Presley, *Saga of Wealth,* 317–18.

23. "He's Half Wildcatter," 1–3.

24. James A. Clark, introduction to Johnny Mitchell, *Energy—And How We Lost It,* xviii.

25. "He's Half Wildcatter," 3.

26. Kirk Mitchell, interview.

27. Jeffrey Share, ed., *The Oil Makers* (College Station: Texas A&M University Press, 2000), 149–50.

28. Untitled press release, Rives, Dyke & Co., May 6, 1965.

29. "He's Half Wildcatter," 2.

30. Kutchin, *How Mitchell Energy,* 405.

31. GPM, interview by Paul Wiesenthal.

32. Kutchin, *How Mitchell Energy,* 405.

33. Roger Galatas and Jim Barlow, *The Woodlands: The Inside Story of Creating a Better Hometown* (Washington, DC: Urban Land Institute, 2004), 11.

34. GPM said the deal was with Louisiana Pacific. Several other employees said it was Georgia Pacific; Kutchin, *How Mitchell Energy,* 405–6.

35. Kutchin, 340.

36. Galatas and Barlow, *Woodlands,* 14.

Chapter 10

1. Sheehy, *Texas Big Rich,* 368.

2. Elizabeth Kolbert, "Dymaxion Man," *New Yorker,* June 9, 2008, accessed May 29, 2016, http://www.newyorker.com/magazine/2008/06/09/dymaxion-man.

3. Kutchin, *How Mitchell Energy,* 437.

4. Kutchin, 438.

5. Jurgen Schmandt, *George P. Mitchell and the Idea of Sustainability* (College Station: Texas A&M University Press, 2010), 8–10.

6. Ware, interview.

7. Scott Mitchell, interview.

8. McQuade, "Good Living in Houston," 111.

9. Todd Mitchell, interview, April 25, 2016.

10. James A. Baker III, interview by the author, January 24, 2017.

11. Reagan Scott Miller, "The Architecture of MacKie and Kamrath" (master's thesis, Rice University, 1993).

12. Todd Mitchell, interview, April 25, 2016.

13. McQuade, "Good Living," 111.

14. Barbi Barbee, "The History of Buffalo Bayou in 1 Minute," Culture Trip, June 30, 2016, accessed January 17, 2017, https://theculturetrip.com/north-america/usa/texas/articles/the-history-of-buffalo-bayou-in-1-minute/.

15. Teresa Tomkins-Walsh, "Houston's Environmental Legacy: Terry Hershey, Community and Action," *Houston History* 10, no. 2 (Spring 2013), 4–5, accessed January 15, 2017, https://houstonhistorymagazine.org/wp-content/uploads/2013/07/Hershey-environmental-legacy.pdf; Frank C. Smith, "It's Time Again to Stop the Bulldozers on the Bayou," *Houston Chronicle*, October 24, 2014, accessed January 17, 2017, http://www.houstonchronicle.com/local/gray-matters/article/Time-once-again-to-stop-the-bulldozers-on-the-5843294.php?cmpid=twitter-premium&t=22a454d7a879b87a02.

16. Scott Mitchell, interview.

17. "About YPO," YPO, accessed January 17, 2017, http://www.ypo.org/about-ypo/.

18. Harrison, "Making of a Billionaire," 70–71.

19. Ware, interview.

20. "Watts Riots," Civil Rights Digital Library, accessed January 24, 2017, http://crdl.usg.edu/events/watts_riots/?Welcome.

21. Schmandt, *George P. Mitchell*, 4–5.

Chapter 11

1. Howard Kiatta, interview by the author, March 4, 2016; Kutchin, *How Mitchell Energy*, 287–88.

2. Purcell, interview.

3. Kutchin, *How Mitchell Energy*, 340.

4. Linda Bomke, interview by the author, May 14, 2016.

5. Kiatta, interview, August 26, 2013.

6. Purcell, interview.

7. Jim Blackburn, interview by the author, March 30, 2016.

8. Smothers, interview.

9. Kiatta, interview, March 4, 2016.

10. Kutchin, *How Mitchell Energy*, 289.

11. Smothers, interview.

12. Kutchin, *How Mitchell Energy*, 198.

13. Bomke, interview.

14. Kutchin, *How Mitchell Energy*, 265.

15. Kiatta, interview, March 4, 2016.

16. Purcell, interview; Kutchin, *How Mitchell Energy*, 19 (production numbers), 652.

17. Boone Pickens, interview by the author, February 19, 2016.

18. Kutchin, *How Mitchell Energy*, 301.

19. *The Associates Press* (company newsletter), September 5, 1971, 3.

20. Kutchin, *How Mitchell Energy*, 201.

21. Jordan Blum, "Shell Oil Helped Ignite the Growth of Houston's Downtown Skyline," *Houston Chronicle*, August 23, 2016, accessed January 24, 2017, http://www.chron.com/local/history/economy-business/article/Shell-Oil-helped-ignite-the-growth-of-Houston-s-9180663.php.

22. Bomke, interview.

23. Kutchin, *How Mitchell Energy*, 199.

24. Purcell, interview.

25. Kutchin, *How Mitchell Energy*, 309.

26. Bomke, interview.

27. Kiatta, interview, March 4, 2016.

28. Purcell, interview.

29. Purcell, interview.

Chapter 12

1. John W. Warnock, "Exploiting Saskatchewan's Potash: Who Benefits?," Canadian Centre for Policy Alternatives, January 2011, accessed January 23, 2017, https://www.policyalternatives.ca/sites/default/files/uploads/publications/Saskatchewan%20office/2011/01/Exploiting%20SK%20Potash%2001–25–11.pdf.

2. Janice Van Dyke Walden, interview by Phil Shook, January 27, 2017.

3. Todd Mitchell, interview, April 25, 2016; Walden, interview.

4. Yergin, *Prize*, 429–30.

5. Cartwright, *Galveston*, 259–61.

6. Cartwright, 299–303.

7. "Logging Basics," *Crain's Petrophysical Handbook*, accessed May 24, 2016, https://www.spec2000.net/01-whatisalog.htm.

8. Cartwright, *Galveston*, 39–53.

9. Purcell, interview.

10. "George P. Mitchell Investing Millions on Galveston Island," *Galveston Daily News*, February 23, 1969, 33, Newspapers.com.

11. Kutchin, *How Mitchell Energy*, 295.

12. Brian Graham, "Mitchell's Galveston Gas Wells Could Hit $2.3 Million Yearly," *Galveston Daily News*, August 12, 1970, 1, Newspapers.com.

13. Sheehy, *Texas Big Rich*, 362–63.

14. "George P. Mitchell Investing Millions," 33.

15. Kutchin, *How Mitchell Energy*, 292; GPM, interview by Paul Wiesenthal, 20; Sheehy, *Texas Big Rich*, 362.

16. GPM, interview by Paul Wiesenthal, 22.

Chapter 13

1. Shaker Khayatt, obituary, *New York Times*, June 12, 2012, accessed April 9, 2016, https://www.legacy.com/obituaries/nytimes/obituary.aspx?n=shaker-a-khayatt&pid=158005708&fhid=11210.

2. Kutchin, *How Mitchell Energy*, 278.

3. Kutchin, 392.

4. Kutchin, 330.

5. Kutchin, 154.

6. Kutchin, 153–54.

7. Kutchin, 282.

8. Steward, *Barnett Shale Play*, 30.

9. "History of Smog," *LA Weekly*, September 22, 2005, accessed April 9, 2016, http://www.laweekly.com/news/history-of-smog-2140714.

10. "History of Smog."

11. GPM, interview by Paul Wiesenthal, 15.

12. Kutchin, *How Mitchell Energy*, 344.

13. "Houston, TX," *Handbook of Texas Online*, Texas State Historical Association, accessed December 7, 2016, https://www.tshaonline.org/handbook/online/articles/hdh03.

14. Kutchin, *How Mitchell Energy*, 407.

15. Lance, "Woodlands," 76.

16. Schmandt, *George P. Mitchell*, 4–5.

Chapter 14

1. Kutchin, *How Mitchell Energy*, 407.

2. Van Caspel, interview, 6.

3. Kutchin, *How Mitchell Energy*, 281.

4. Lance, "Woodlands," 118.

5. Lance, 80.

6. Lance, 80.

7. Kutchin, *How Mitchell Energy*, 358.

8. Galatas and Barlow, *Woodlands*, 14–15.

9. Galatas and Barlow, 14–15.

10. Ann Forsyth, *Reforming Suburbia: The Planned Communities of Irvine, Columbia, and The Woodlands* (Berkeley: University of California Press, 2005), 168.

11. Galatas and Barlow, *Woodlands*, 16.

12. Kutchin, *How Mitchell Energy*, 350.

13. Galatas and Barlow, *Woodlands*, 16.

14. Galatas and Barlow, 33.

15. "Socially Responsible Entrepreneur of the Year," undated press release among GPM's papers.

16. Robert Heineman, interview by the author, March 4, 2016.

17. Robert Heineman, interview.

18. Robert Heineman, interview.

19. Roger Galatas, "General Outline Covering Lessons Learned from the Development of The Woodlands and Other New Town Developments in the Greater Houston Region" (unpublished manuscript, October 2, 2015), 1.

20. Galatas, 5.

21. Galatas, 5.

22. Lance, "Woodlands," 118.

23. Heineman, interview.

24. Scott Mitchell, interview.

25. Heineman, interview.

26. Scott Mitchell, interview.

27. Schmandt, *George P. Mitchell*, 6–7.

28. Kutchin, *How Mitchell Energy*, 280.

29. Kutchin, 280.

30. Kutchin, 604–5.

31. Scott Mitchell, interview.

32. Scott Mitchell, interview.

33. Scott Mitchell, interview.

34. Scott Mitchell, interview.

35. Heineman, interview.

36. Ware, interview.

37. Scott Mitchell, interview.

Chapter 15

1. Galatas and Barlow, *Woodlands*, 47.

2. Blackburn, interview.

3. Kutchin, *How Mitchell Energy*, 352.

4. Galatas and Barlow, *Woodlands*, 55.

5. Kutchin, *How Mitchell Energy*, 325.

6. Ware, interview.

7. Heineman, interview.

8. Galatas and Barlow, *Woodlands*, 47.

9. Heineman, interview.

10. Ware, interview.

11. Galatas and Barlow, *Woodlands*, 48.

12. Galatas, "General Outline Covering Lessons Learned," 9.

13. Ware, interview.

14. Galatas and Barlow, *Woodlands*, 48.

15. Galatas and Barlow, 47.

16. Galatas and Barlow, 49.

17. Blackburn, interview.

18. Lance, "Woodlands," 82.

19. Galatas and Barlow, *Woodlands*, 51.

20. Coulson Tough, interview by the author, March 3, 2016.

21. Galatas and Barlow, *Woodlands*, 56.

22. Montgomery County, Texas, tax records; Sheridan Lorenz, email to author, January 2, 2017.

23. Tough, interview.

24. Tough, interview.

25. Galatas and Barlow, *Woodlands*, 30–33.

26. Lance, "Woodlands," 76.

27. Lance, 76.

28. Schmandt, *George P. Mitchell*, 7.

Chapter 16

1. "George P. Mitchell Investing Millions," 33.

2. Scott Mitchell, interview.

3. Sheehy, *Texas Big Rich*, 364.

4. "Mike Mitchell, for Whom Campus Was Named, Dies," 1.

5. Christie Mitchell to Maria Mitchell Ballantyne, undated (about 1962).

6. Christie Mitchell, "To the End."

7. "Christie Mitchell Dies," *Galveston Daily News*, March 1, 1980, 1, Newspapers.com.

8. Johnny Mitchell, "Beachcomber's Inspiration Came from His Love of People of All Types," *Galveston Daily News*, March 17, 1980, 4, Newspapers.com; "Discrimination Lawsuit Filed," *Galveston Daily News*, January 7, 1993, 2, Newspapers.com.

9. Sheehy, *Texas Big Rich*, 365.

10. GPM, interview by Paul Wiesenthal.

11. "Strand's Historic League-Blum Building Restored to Its Former Elegance," *Galveston Daily News*, May 13, 1979, 1C.

12. "Strand Historic District, T. Jefferson League Building," Historic American Buildings Survey, Heritage Conservation and Recreation Service, US Department of the Interior, HABS No. TX-3296 F, 1–3, accessed February 4, 2017, http://cdn.loc.gov/master/pnp/habshaer/tx/tx0000/tx0047/data/tx0047data.pdf; "About GHF," Galveston Historical Foundation, accessed February 4, 2017, http://www.galvestonhistory.org/about-ghf/about-ghf.

13. Tough, interview.

14. Todd Mitchell, interview, April 25, 2017.

15. The account of the Mitchells' marital difficulties is based on interviews with family members and former employees, many of whom asked that their names not be used. I have withheld the identity of GPM's paramour, who is deceased, out of respect for her family, who were apparently unaware of the affair.

16. Sheehy, *Texas Big Rich*, 365.

17. Sheehy, 365.

18. Todd Mitchell, interview, April 25, 2016.

19. Bomke, interview.

20. "Mardi Gras History," Galveston.com, accessed February 4, 2017, http://www.galveston.com/mardigrashistory/.

21. Ware, interview.

22. Kutchin, *How Mitchell Energy*, 145.

23. Kirk Mitchell, interview.

24. Scott Mitchell, interview.

25. Kutchin, *How Mitchell Energy*, 440–41.

26. Bomke, interview.

27. Ware, interview.

28. Kutchin, *How Mitchell Energy*, 441.

29. James A. Baker III and Steve Fiffer, *"Work Hard, Study . . . and Keep Out of Politics!"* (Evanston, IL: Northwestern University Press, 2006).

30. Baker, interview.

31. Dale Robertson, "Bayou City Has Gained Its Status as a Tennis Hotbed," *Houston Chronicle*, April 10, 2005, accessed December 5, 2016, http://www.chron.com/sports/article/Bayou-City-has-regained-its-status-as-a-tennis-1942802.php.

32. Lena Williams, "Gladys Heldman, 81, a Leader in Promoting Womens Tennis," *New York Times*, June 25, 2003, accessed December 5, 2016, http://www.nytimes.com/2003/06/25/sports/gladys-heldman-81-a-leader-in-promoting-womens-tennis.html; Dale Robertson, "King Won 'Battle of the Sexes' and Served Notice for Equality," *Houston Chronicle*, May 28, 2016, accessed May 28, 2016, http://www.houstonchronicle.com/local/history/major-stories-events/article/The-night-King-won-Battle-of-the-Sexes-and-7951577.php.

33. Williams, "Gladys Heldman."

34. Robertson, "Bayou City Has Gained Its Status"; Lorenz, interview, March 4, 2016.

35. "George P. Mitchell," Texas Tennis Museum and Hall of Fame," accessed May 30, 2016, https://www.texastennismuseum.org/mitchell.

36. Mark Seal, *Raising the Bar: The Life and Work of Gerald D. Hines* (Bainbridge Island, WA: Fenwich Publishing, 2016), 239.

37. Gerald Hines, interview by the author, May 9, 2016.

38. Dennis L. Meadows, preface to *Limits to Growth: The 30-Year Update*, by Donella Meadows, Jorgen Randers, and Dennis L. Meadows (White River Junction, VT: Chelsea Green, 2004), ix–x.

39. Schmandt, *George P. Mitchell*, 30.

40. Schmandt, 43.

41. Mary Davis Sure, "George P. Mitchell: Recalling Twenty Years of Global Caretaking," *The Woodlands Forum*, Winter 1993–1994, 3.

42. Sure, 8.

43. Schmandt, *George P. Mitchell*, 44.

44. Harrison, "Making of a Billionaire," 70.

45. Gary Taylor, "Open Forum," *Houston Metropolitan*, February 1990, 29–33; Tony Lentini, interview by the author, May 1, 2016.

46. GPM, interview by Paul Wiesenthal, 23, 35.

Chapter 17

1. Robert R. Wilson, "The Tevatron," *Physics Today*, October 1977, 23, accessed March 5, 2017, http://dx.doi.org/10.1063/1.3037746.

2. The point of the ever-bigger machines is to produce more energetic collisions, which reveal the more massive particles or tear apart clingy particles into their constituents. However, as the beams approach the speed of light, the particles get more and more massive, and it requires ever-increasing amounts of energy for that next increment of speed. The beams also want to go straight, but a circle allows them to accelerate in a finite space, gaining speed as they travel around the circle. It takes energy to bend the beams, and there's a practical limit to the curvature of the beam that won't compromise energy. Europe's Large Hadron Collider, for example, uses lower energy (to fit in the existing tunnel) and denser beams to increase the chances of collision. It turned out that 13 TeV (teraelectronvolts) and a higher collision rate were enough to detect the Higgs boson.

3. Michael Riordan, Lillian Hoddeson, and Adrienne W. Kolb, *Tunnel Visions* (Chicago: University of Chicago Press, 2015), ix–37.

4. Trevor Quirk, "How Texas Lost the World's Largest Super Collider," *Texas Monthly*, October 23, 2013, accessed February 28, 2017, http://www.texasmonthly.com/articles/how-texas-lost-the-worlds-largest-super-collider/.

5. "Fermilab History and Archives Project," Fermilab, accessed March 1, 2017, http://history.fnal.gov/transition.html.

6. Tom Adair, interview by the author, January 19, 2016; Schmandt, *George P. Mitchell*, 58–59.

7. GPM, interview by Paul Wiesenthal, 20.

8. Todd Mitchell, interview, April 25, 2016.

9. Wilson, "Tevatron," 23.

10. John Matson, "New Particle Resembling Long-Sought Higgs Boson Uncovered at Large Hadron Collider," *Scientific American*, July 4, 2012, accessed March 1, 2017, https://www.scientificamerican.com/article/higgs-cern-lhc-discovery/.

11. Ware, interview.

12. Quirk, "How Texas Lost."

13. Jim Merithew, "For Sale: $20 Million Particle Accelerator, Never Used," *Wired*, September 9, 2009, accessed February 28, 2017, https://www.wired.com/2009/09/super-collider-gallery/.

14. Riordan, Hoddeson, and Kolb, *Tunnel Visions*, 51–53.

15. Jack Z. Smith, "High-Energy Guru," *Fort Worth Star-Telegram*, July 17, 1991, D1.

16. David Appell, "The Supercollider That Never Was," *Scientific American*, October 15, 2013, accessed February 28, 2017, https://www.scientificamerican.com/article/the-supercollider-that-never-was/.

Chapter 18

1. Kutchin, *How Mitchell Energy*, 331.

2. Michael Rounds, "Oil Patch Numbers Paint Grim Picture," *Houston Chronicle*, August 17, 1986, sec. 5, 14.

3. Kutchin, *How Mitchell Energy*, 283.

4. GPM, interview by Paul Wiesenthal, 14.

5. Kutchin, *How Mitchell Energy*, 333.

6. Toni Mack, "Staying Power," *Forbes*, January 21, 1991, 81, Business Source Complete (9101210356).

7. Mack, 81.

8. Homer Hershey, interview by the author, May 5, 2016.

9. Mitchell Energy and Development Corp., 10-K annual report for the fiscal year ended January 31, 1994, filed with the US Securities and Exchange Commission, April 26, 1994, https://www.sec.gov/Archives/edgar/data/311995/0000950129–94–000319.txt.

10. Bloomberg stock data.

11. Ware, interview.

12. Robert Johnson and Allanna Sullivan, "Mitchell Picks Stevens to Be Its President, Operations Chief," *Wall Street Journal*, December 13, 1993, B6, ProQuest (02761849).

13. Tim Beims, "Storied Past, Rosy Future Keep Mitchell Energy Hot on Natural Gas," *American Oil & Gas Reporter*, August 2000, 58, copy from GPM's personal papers.

14. Lorenz, email.

15. Van Caspel, interview, 8.

16. Johnson and Sullivan, "Mitchell Picks Stevens."

17. Stevens, through a spokesperson for EOG Resources, declined an interview.

18. Lentini, interview, March 1, 2016.

19. Lentini, interview, March 1, 2016.

20. Lentini, interview, March 1, 2016.

21. Steward, "George P. Mitchell," 60.

22. Vikram Rao, *Shale Oil and Gas: The Promise and the Peril* (Research Triangle Park, NC: Research Triangle Institute Press, 2015), 3–6.

23. Dan Steward, interview by the author, January 4, 2016.

24. "Project Gasbuggy Tests Nuclear 'Fracking,'" American Oil & Gas Historical Society, undated, https://aoghs.org/technology/project-gasbuggy/.

25. Michael MacRae, "Fracking: A Look Back," American Society of Mechanical Engineers, December 2012, accessed January 6, 2016, https://www.asme.org/engineering-topics/articles/fossil-power/fracking-a-look-back.

26. Luke Yoquinto, "The Truth about Guar Gum," *LiveScience*, July 27, 2012, accessed January 7, 2016, http://www.livescience.com/36580-guar-gum-weight-loss-cost.html.

27. Steward, interview, January 4, 2016.

28. The reason for the marginal economics had as much to do with accounting rules as technology. The Securities and Exchange Commission wouldn't allow Mitchell Energy to book higher reserves from the Barnett because there was nothing to compare it to. Essentially, the agency didn't want the company making up overoptimistic estimates, so it required more conservative accounting.

29. Kiatta, interview, March 4, 2016.

Chapter 19

1. Janin Friend, "Flaming Justice," *Texas Business*, January 1997, 51.

2. Friend, 51.

3. Michael Davis, "Mitchell, Residents Square Off," *Houston Chronicle*, July 26, 1996, 1E, clipping from GPM's personal papers.

4. Hershey, interview, May 5, 2016.

5. Gordon E. Taylor (former Mitchell Energy engineer) to GPM, June 2, 1997, from GPM's private papers.

6. Russell Gold, *The Boom: How Fracking Ignited the American Energy Revolution and Changed the World* (New York: Simon & Schuster, 2014), 97–101.

7. Jay Ewing, interview by the author, January 27, 2016.

8. Kutchin, *How Mitchell Energy*, 452.

9. "Socially Responsible Entrepreneur of the Year," undated press release.

10. Davis, "Mitchell, Residents Square Off."

11. Friend, "Flaming Justice," 48.

12. Ewing, interview.

13. Friend, "Flaming Justice, 49.

14. Friend, 50.

15. "Smelly Verdict—II" (unsigned editorial), *Wall Street Journal*, June 12, 1996, copy from GPM's private papers.

16. Friend, "Flaming Justice," 51.

17. Friend, 51.

18. Friend, 52.

19. "Mitchell Energy Wins Significant Bailey Water Well Verdict" (Mitchell Energy press release), PR Newswire, May 28, 1997.

20. Jack O'Neill, interview by the author, January 11, 2018.

21. Hershey, interview.

22. Michael Davis, "Mitchell Wins Suit over Water," *Houston Chronicle*, May 29, 1997, 1C, clipping from GPM's personal papers.

23. "Mitchell Energy Wins."

24. "Texas Court of Appeals Reverses Pollution Ruling," *Wall Street Journal*, November 14, 1997, B6, clipping from GPM's personal papers.

25. Gold, *Boom*, 109.

26. Paul Monies, "Landowners File Suit against Devon Energy." *Oklahoman*, June 9, 2004, accessed May 8, 2016, https://newsok.com/article/1906186/landowners-file-suit-against-devon-energy.

27. Gold, *Boom*, 110–11.

28. Ralph Bivins, "Mitchell Selling The Woodlands," *Houston Chronicle*, June 13, 1997, 1A.

Chapter 20

1. "Mitchell Energy & Development: Gas-Supply Pact with Unit of Occidental to Terminate," *Wall Street Journal*, June 19, 1995, C15, ProQuest (03568660).

2. Hershey, interview.

3. Nick Steinsberger, interview by the author, February 19, 2016.

4. "Cushing, OK WTI Spot Price FOB."

5. Steward, interview, January 8, 2018.

6. Gold, *Boom*, 117.

7. Steward, interview, January 4, 2016.

8. Gregory Zuckerman, *The Frackers: The Outrageous Insider Story of the New Billionaire Wildcatters* (New York: Penguin Group, 2013), 89.

9. Kent Bowker, interview by the author, January 8, 2016.

Chapter 21

1. The assessment of Stevens's views and the interaction between Stevens and GPM is compiled from a number of former employees who had regular contact with both men at Mitchell Energy.

2. Sure, "George P. Mitchell," 1.

3. Todd Mitchell, interview, March 22, 2016.

4. Todd Mitchell, interview, April 25, 2016.

5. Todd Mitchell, interview, April 25, 2016.

6. Todd Mitchell, interview, April 25, 2016.

7. Mitchell Energy and Development, 10-K annual report for the fiscal year ended January 31, 1999, filed with the US Securities and Exchange Commission, April 16, 1999, https://www.sec.gov/Archives/edgar/data/311995/0000950129–98–001454.txt.

8. Mitchell Energy and Development, 10-K annual report for the fiscal year ended January 31, 1999.

9. Bloomberg stock data.

10. "Socially Responsible Entrepreneur of the Year," undated press release.

11. Scott Mitchell, interview.

Chapter 22

1. "Johnny Michael Mitchell, Sr.," obituary, Find a Grave, accessed December 13, 2014, https://www.findagrave.com/cgi-bin/fg.cgi?page=gr&GRid=30099318.

2. "Mitchell Energy Sells Properties," *Wall Street Journal*, July 2, 1999, 1, ProQuest (39868330).

3. "Mitchell Energy & Development Corp.: Advisers Are Hired to Look into Strategic Alternatives," *Wall Street Journal*, October 7, 1999, B7, ProQuest (398789439).

4. Spiros Vassilakis, interview by the author, March 16, 2016.

5. Todd Mitchell, interview, April 25, 2016.

6. Larry Nichols, interview by the author, February 25, 2016.

7. Todd Mitchell, interview, March 22, 2016.

8. "Mitchell Energy & Development Corp.: Firm to Stay Independent, Expects to Beat Forecasts," *Wall Street Journal*, April 7, 2000, A6, ProQuest (398712425).

9. Ware, interview.

10. Kutchin, *How Mitchell Energy*, 285–86.

11. Mitchell Energy and Development, 10-K annual report for the year ended December 31, 2000, filed with the US Securities and Exchange Commission, March 9, 2001, https://www.sec.gov/Archives/edgar/data/311995/000095012901001352/h84720e10-k405.txt.

12. Robin Sidel and Chip Cummins, "Devon Energy to Acquire Mitchell in

$3.1 Billion Cash-and-Stock Deal," *Wall Street Journal*, August 14, 2001, accessed May 1, 2016, https://www.wsj.com/articles/SB997742296816559591.

13. Sidel and Cummins.

14. "Henry Hub Natural Gas Spot Price," US Energy Information Administration, accessed May 9, 2016, https://www.eia.gov/dnav/ng/hist/rngwhhdm.htm.

15. "EOG Resources Elects Bill Stevens and Leighton Steward to Board of Directors," company press release, July 2, 2004.

Chapter 23

1. Kutchin, *How Mitchell Energy*, 281.
2. Todd Mitchell, interview, April 25, 2016
3. Joe Greenberg, interview by the author, April 8, 2016.
4. Jennifer McCarthy, interview by the author, April 8, 2016.
5. Todd Mitchell, interview, April 25, 2016.
6. Greenberg, interview.
7. Scott Mitchell, interview by the author, April 25, 2016.
8. Wendy Mitchell, interview by the author, April 25, 2016.
9. Greenberg, interview.
10. "Henry Hub Natural Gas Spot Price."
11. McCarthy, interview.

Chapter 24

1. "Hawking Condemns US Cancellation of SSC," *Nature* 392, no. 118 (March 12, 1998), accessed September 17, 2016, https://www.nature.com/articles/32264.

2. Eric Berger, "Money Made in Oil, Mitchell Dreamed of the Stars," *Houston Chronicle*, August 3, 2013, accessed November 2, 2014, http://www.houstonchronicle.com/news/nation-world/article/Money-made-in-oil-Mitchell-dreamed-of-the-stars-4704432.php.

3. Ed Fry, interview by the author, February 15, 2016.

4. "Prestigious Stephen Hawking Chair in Physics Goes to Christopher Pope," Texas A&M University press release, October 28, 2002, accessed February 29, 2016, http://www.science.tamu.edu/news/story.php?story_ID=13#.V_LQWJMrJ24.

5. "Mitchell Tennis Center," Texas A&M University, Quick Facts, accessed December 5, 2016, http://www.12thman.com/facilities/?id=13; "George P. Mitchell '40," Texas A&M University Corps of Cadets Hall of Honor, accessed February 15, 2016, http://corps.tamu.edu/portfolio-items/george-mitchell/.

6. Christopher Pope, interview by the author, February 29, 2016.

7. Pope, interview.

8. Fry, interview.

9. Among those who attended the meeting, there's some uncertainty as to whether Hawking brought the idea up first.

10. Fry, interview. The article appeared in the January 21, 2005, issue (vol. 307), 340.

11. Sally Lehrman, "He'll Pay for That," *Scientific American*, July 1, 2005, accessed October 4, 2016, http://www.scientificamerican.com/article/hell-pay-for-that/.

12. Fry, interview.

13. Peter McIntyre, interview by the author, January 7, 2016.

14. Joe Newton, interview by the author, March 21, 2016.

15. This account is taken from a recollection by Joe Newton, published on the *Texas A&M Science Blog* on January 29, 2014. Fry agrees with most of Newton's recollection but insists that it was George, not Gates, who offered the $2.5 million over ten years.

16. Shana Hutchins, "Life Forces and Legacies," *Texas A&M Science Blog*, January 29, 2014, accessed October 4, 2016, https://tamuscience.com/2014/01/29/life-forces-and-legacies/.

17. Steve Koppes, "Wendy Freedman, World-Leading Astronomer, Joins UChicago Faculty," *University of Chicago News*, August 7, 2014, accessed April 5, 2016, https://news.uchicago.edu/article/2014/08/07/wendy-freedman-world-leading-astronomer-joins-uchicago-faculty.

18. "Quick Facts," Giant Magellan Telescope, accessed April 7, 2016, http://www.gmto.org/overview/quick-facts/.

19. "What Is GMT?," Giant Magellan Telescope, accessed April 7, 2016, http://www.gmto.org/overview/.

20. "George Mitchell Commits $25 Million to Giant Magellan Telescope," Giant Magellan Telescope, May 1, 2011, accessed April 9, 2016, https://www.gmto.org/2011/05/george-mitchell-commits-25-million-to-giant-magellan-telescope/.

21. Ware, interview.

Chapter 25

1. Claudio Soto, interview by the author, March 17, 2016.

2. Lorenz, interview, March 14, 2016.

3. Herb Magley, interview by the author, March 10, 2016. Magley declined to share details of his conversations with GPM regarding their wives' illnesses.

4. "Cynthia Woods Mitchell Dies at Age 87," KHOU.com, December 27, 2009, accessed March 25, 2017, https://www.khou.com/article/news/cynthia-woods-mitchell-dies-at-age-87/285–342585282.

5. Ware, interview.

6. Lorenz, interview, March 14, 2016.

7. Kirk Mitchell, interview.

8. Colorado Oil and Gas Conservation Commission complaint reports, received May 23, 2008, http://cogcc.state.co.us/cogis/ComplaintReport.asp?doc_num=200190138; Steve Everley, "From Flaming Faucet to Flaming Hose: The Continuing Fraud of Gasland," Energy in Depth, July 7, 2013, https://www.energyindepth.org/national/the-continuing-fraud-of-gasland/.

9. Lawrence Wright, "The Glut Economy," *New Yorker*, January 1, 2018, 48.

10. Lorenz, email.

11. GPM to Cynthia Woods Mitchell, undated latter provided by Sheridan Lorenz.

12. Stephen Hawking, audio recording provided by Sheridan Lorenz of GPM eulogy played at a memorial service at the Cynthia Woods Mitchell Pavilion in The Woodlands, August 8, 2013.

13. M. A. Caravageli, MD, to GPM, February 21, 1991.

Chapter 26

1. James Taylor, "Fracking, Lower Gasoline Prices Returned $1,000 to Household Budgets Last Year," Forbes.com, February 3, 2017, accessed March 15, 2017, https://www.forbes.com/sites/jamestaylor/2017/02/03/fracking-lower-gasoline-prices-returned-1000-to-household-budgets-last-year/#73f8e66030ce.

2. Ed Crooks, "US Fracking Shown to Benefit Local Households," FT.com, January 2, 2017, accessed March 29, 2017, https://www.ft.com/content/cbee8072-c973–11e6–8f29–9445cac8966f.

3. Robert Rapier, "Fracking Saves Americans $180 Billion Annually on Gasoline," Forbes.com, June 2, 2017, accessed June 2, 2017. https://www.forbes.com/sites/rrapier/2017/06/02/fracking-saves-consumers-180-billion-annually-on-gasoline/2/#2e85eef32e88.

4. Sarah McFarlane and Summer Said, "Tables Turned: Saudi Arabia Hunts for Oil Assets in the US," *Wall Street Journal*, December 20, 2017, https://www.wsj.com/article_email/saudi-arabia-searches-for-shale-oil-deal-1513782415-lMyQjAxMTI3MDI2MTEyMjEzWj/.

5. John Bringardner, "The Fracking Boom and the Fall of a Texas Utility," *New Yorker*, May 2, 2014.

6. Justin L. Rubinstein and Alireza Babaie Mahani, "Myths and Facts on Wastewater Injection, Hydraulic Fracturing, Enhanced Oil Recovery, and Induced Seismicity," *Seismological Research Letters* 86, no. 4 (July/August 2015): 1–8, doi:10.1785/0220150067.

7. Rubinstein and Mahani, 1–8.

8. Kent A. Bowker, "Barnett Shale Gas Production, Fort Worth Basin: Issues

and Discussion," *AAPG Bulletin* 91, no. 4 (April 2007): 523–33.

9. "States Take Wait and See Approach on Fracking Regulation," Congress.org, July 9, 2015, accessed May 24, 2017, http://congress.org/2015/07/09/states-take-wait-and-see-approach-on-fracking-regulation/.

10. Lance, "Woodlands," 81.

11. "George Mitchell: The Energy Man," *Greek Accent*, September/October 1986, 15.

12. "Population by Village," The Woodlands Township, April 2016, accessed March 30, 2017, https://www.thewoodlandstownship-tx.gov/documentcenter/view/8902.

13. GPM to the residents of The Woodlands, June 16, 1997.

14. John Schwartz, James Glanz, and Andrew W. Lehren, "Builders Said Their Homes Were out of a Flood Zone. Then Harvey Came," *New York Times*, December 3, 2017, A1, https://www.nytimes.com/2017/12/02/us/houston-flood-zone-hurricane-harvey.html?_r=0.

15. "The Woodlands, TX Demographics," areavibes.com., accessed March 30, 2017, http://www.areavibes.com/the+woodlands-tx/demographics/.

16. "The Woodlands Home Prices & Values," Zillow.com, accessed March 30, 2017, https://www.zillow.com/the-woodlands-tx/home-values/; "Houston Metro Home Prices & Values," Zillow.com, accessed March 30, 2017, http://www.zillow.com/houston-metro-tx_r394692/home-values/.

17. "The Woodlands CDP, Texas," US Census Bureau, accessed March 30, 2016, https://www.census.gov/quickfacts/table/HSG010215/4872656,00; "About Affordable Housing in The Woodlands," affordablehousingonline.com, accessed March 30, 2017, https://affordablehousingonline.com/housing-search/Texas/The-Woodlands.

18. 2007 American Community Survey, 1-Year Estimates, US Census Bureau, accessed March 31, 2017, https://factfinder.census.gov/faces/tableservices/jsf/pages/productview.xhtml?src=bkmk; 2007 American Community Survey.

19. Galatas, "General Outline Covering Lessons Learned," 1.

20. Kamran Rosen, "Best Places to Live in Texas," Nerd Wallet, June 15, 2015, accessed March 31, 2017, https://www.nerdwallet.com/blog/mortgages/best-places-to-live-in-texas-2015/.

21. Harrison, "Making of a Billionaire," 70.

22. Kiatta, interview, March 3, 2016.

Selected Bibliography

Articles

Appell, David. "The Supercollider That Never Was." *Scientific American*, October 15, 2013. Accessed February 28, 2017. https://www.scientificamerican.com/article/the-supercollider-that-never-was/.

Beims, Tim. "Storied Past, Rosy Future Keep Mitchell Energy Hot on Natural Gas." *American Oil & Gas Reporter*, August 2000, 58.

Berger, Eric. "Money Made in Oil, Mitchell Dreamed of the Stars." *Houston Chronicle*, August 3, 2013. Accessed November 2, 2014. http://www.houstonchronicle.com/news/nationworld/article/Money-made-in-oil-Mitchell-dreamed-of-the-stars-4704432.php.

Bivins, Ralph, and Andrea Ahles. "Sale Makes Some Residents Uneasy." *Houston Chronicle*, June 14, 1997, 1C.

Blum, Jordan. "Shell Oil Helped Ignite the Growth of Houston's Downtown Skyline." *Houston Chronicle*, August 23, 2016. Accessed January 24, 2017. http://www.chron.com/local/history/economy-business/article/Shell-Oil-helped-ignite-the-growth-of-Houston-s-9180663.php.

Businessweek. "He's Half Wildcatter, Half Big Businessman." November 24, 1964.

Corsicana Daily Sun. "Big Oil, Gas Deal Reported." January 1, 1963, 8. Newspapers.com.

Crooks, Ed. "US Fracking Shown to Benefit Local Households." FT.com, January 2, 2017. Accessed March 29, 2017. https://www.ft.com/content/cbee8072-c973–11e6–8f29–9445cac8966f.

Davis, Michael. "Mitchell, Residents Square Off." *Houston Chronicle*, July 26, 1996, 1E.

______. "Mitchell Wins Suit over Water." *Houston Chronicle*, May 29, 1997, 1C.

______. "Questions Raised about Mitchell's Next Steps." *Houston Chronicle*, June 14, 1997, 3C.

Draper, Robert. "Big Fish." *Texas Monthly*, May 1997. Accessed August 10, 2015. http://www.texasmonthly.com/articles/big-fish/.

Elder, Laura. "Billionaire Looks to Conservation." *Galveston Daily News*, January 1, 2005. Newspapers.com.

Feldman, Claudia. "Low-Key Billionaire." *Houston Chronicle*, *Texas* magazine supplement, June 22, 2003, 9–13.

Friend, Janin. "Flaming Justice." *Texas Business*, January 1997, 48–51.

Galehouse, Maggie. "Last Hurrah for the Houston Club's Rusk Location." *Houston Chronicle*, December 30, 2012. Accessed January 10, 2017. http://www.houstonchronicle.com/news/houston-texas/houston/article/Last-hurrah-for-the-Houston-Club-s-Rusk-location-4156030.php.

Galveston Daily News. "Barker Enjoins Places on Beach." July 31, 1929. Newspapers.com.

———. "Christie Mitchell Dies." March 1, 1980, 1. Newspapers.com.

———. "Discrimination Lawsuit Filed." January 7, 1993, 2. Newspapers.com.
———. "George P. Mitchell Investing Millions on Galveston Island." February 23, 1969, 33. Newspapers.com.
———. "Mike Mitchell, for Whom Campus Was Named, Dies." December 26, 1970, 1. Newspapers.com.
———. "Snug Harbor Fire Suit Seeking $130,000 Damages." December 5, 1947, 20. Newspapers.com.
———. "Three Men Give Themselves Up." November 3, 1914, 3. Newspapers.com.
———. "Trial of Large Damage Claim Opens in Court." April 19, 1949, 1. Newspapers.com.
Gesalman, Anne Belli. "George Mitchell, a Work Ethic Inspired by the Past Extends to the Future." *Dallas Morning News*, February 9, 1997, 4E.
Gold, Russell. "Boom Town: Drilling for Natural Gas Faces a Sizable Hurdle: Fort Worth." *Wall Street Journal*, April 29, 2005. A1.
______. "The Man Who Pioneered the Shale-Gas Revolution." *Wall Street Journal*, October 23, 2012. Newspapers.com.
Graham, Brian. "Mitchell's Galveston Gas Wells Could Hit $2.3 Million Yearly." *Galveston Daily News*, August 12, 1970, 1. Newspapers.com.
Gray, Lisa. "Bayous for Dummies." *Houston Chronicle*, October 17, 2010. Accessed January 17, 2017. http://www.chron.com/life/article/Bayous-for-Dummies-1706294.php.
Harrison, Joanne. "The Making of a Billionaire." *Houston City Magazine*, March 1981, 52–71.
"Hawking Condemns US Cancellation of SSC." *Nature* 392, no. 118 (March 12, 1998). Accessed September 17, 2016. https://www.nature.com/articles/32264.
Herz, Lillian E. "Christie Mitchell Spreads Word of the Isle's Charms." *Galveston News Tribune*, August 23, 1964, 3. Newspapers.com.
Hinton, Diana Davids. "The Seventeen-Year Overnight Wonder: George Mitchell and Unlocking the Barnett Shale." *Journal of American History*, June 2012, 229–35. Accessed November 12, 2015. doi:10.1093/jahist/jas064.
Hirtenstein, Anna. "Record Green Power Installations Beat Fossil Fuel for First Time." *Bloomberg*, October 25, 2016. https://www.bloomberg.com/news/articles/2016–10–25/record-green-power-installations-beat-fossil-fuel-for-first-time.
Houston Post. "Cheaper Rents on Main Street." March 14, 1915, 24. Newspapers.com.
———. "Mitchell and Chiacos Brothers Were Released under $2,000 Bonds." September 1, 1914, 9. Newspapers.com.
———. "Several Indicted Men Gave Bonds on Monday." November 3, 1914, 9. Newspapers.com.
———. "Two Men Who Lost Their Lives When Burning Building Was Wrecked by an Explosion Have Been Identified as Joe Nader and Maynard Upton." August 24, 1914, 1. Newspapers.com.
———. "Widow of Victim of Main Street Fire Gets Check from Magnolia Park Land Co., in Settlement of Death Claim." November 1, 1914, 25. Newspapers.com.
Johnson, Robert, and Allanna Sullivan. "Mitchell Picks Stevens to Be Its President, Operations Chief." *Wall Street Journal*, December 13, 1993, B6. ProQuest (02761849).

Kleinfeld, N. R. "Fraying at the Edges." *New York Times*, May 1, 2016, 1–11 S.
Kolbert, Elizabeth. "Dymaxion Man." *New Yorker*, June 9, 2008. Accessed May 29, 2016. http://www.newyorker.com/magazine/2008/06/09/dymaxion-man.
Lance, Mary. "The Woodlands: Is It Working?" *Southwest Airlines Magazine*, August 1982, 76–118.
LA Weekly. "History of Smog." September 22, 2005. Accessed April 9, 2016. http://www.laweekly.com/news/history-of-smog-2140714.
Lehrman, Sally. "He'll Pay for That." *Scientific American*, July 1, 2005. Accessed October 4, 2016. http://www.scientificamerican.com/article/hell-pay-for-that/.
Lynch, Dudley. "The Frustration Fields Gang Rides Again." *Dallas Morning News*, *Southwest Scene* magazine, December 10, 1972.
Mack, Toni. "Staying Power." *Forbes*, January 21, 1991, 81. Business Source Complete (9101210356).
Matson, John. "New Particle Resembling Long-Sought Higgs Boson Uncovered at Large Hadron Collider." *Scientific American*, July 4, 2012. Accessed March 1, 2017. https://www.scientificamerican.com/article/higgs-cern-lhc-discovery/.
Maykuth, Andrew. "Tapping Shale, Seeking Sustainability. A Rare Oilman." *Philadelphia Inquirer*, August 29, 2010. Accessed January 3, 2016. http://www.philly.com/philly/news/special_packages/inquirer/marcellus-shale/20100829_Tapping_shale__seeking_sustainability__A_Rare_Oilman.html.
McQuade, Walter. "Good Living in Houston: At Home beside the Bayou." *Fortune*, July 1, 1966, 111.
Merithew, Jim. "For Sale: $20 Million Particle Accelerator, Never Used." *Wired*, September 9, 2009. Accessed February 28, 2017. https://www.wired.com/2009/09/super-collider-gallery/.
Mitchell, Christie. "Mitchell Family to Get Together." *Galveston Daily News*, March 27, 1977, 61. Newspapers.com.
_______. "To the End, Mike Had a Will to Live." *Galveston Daily News*, December 27, 1970, 10. Newspapers.com.
Mitchell, George P. "Orphaned Gas: Bridging the Energy Gap." *World Energy* 10, no. 3 (2007): 108–11.
Mitchell, Johnny. "Beachcomber's Inspiration Came from His Love of People of All Types." *Galveston Daily News*, March 17, 1980, 4. Newspapers.com.
Monies, Paul. "Landowners File Suit against Devon Energy." *Oklahoman*, June 9, 2004. Newspapers.com.
New York Times. "Shaker Khayatt." Obituary. June 12, 2012. Newspapers.com.
Petmezas, Socrates. "History of Modern Greece." In *About Greece*, edited by G. Metaxas, 7–41. Athens, 2002.
Peyton, Lindsay. "Roger Galatas Helps Link the Past, Present and Future of The Woodlands." *Houston Chronicle*, October 17, 2014. Accessed June 4, 2016. http://www.chron.com/neighborhood/woodlands/news/article/Roger-Galatas-helps-link-the-past-present-and-5821847.php.
Quirk, Trevor. "How Texas Lost the World's Largest Super Collider." *Texas Monthly*, October 23, 2013. Accessed February 28, 2017. http://www.texasmonthly.com/articles/how-texas-lost-the-worlds-largest-super-collider/.
Radcliffe, Jennifer. "Mitchell's $35 Million Gift to A&M Physics." *Houston Chronicle*, November 3, 2005. Accessed March 16, 2016. http://www.chron.com/news/

houston-texas/article/Mitchell-s-35-million-a-gift-to-A-M-physics-1580309.php.
Robertson, Dale. "Bayou City Has Gained Its Status as a Tennis Hotbed." *Houston Chronicle*, April 10, 2005. Accessed December 5, 2016. http://www.chron.com/sports/article/Bayou-City-has-regained-its-status-as-a-tennis-1942802.php.
______. "King Won 'Battle of the Sexes' and Served Notice for Equality." *Houston Chronicle*, May 28, 2016. Accessed May 28, 2016. https://www.chron.com/local/history/major-stories-events/article/The-night-King-won-Battle-of-the-Sexes-and-7951577.php.
Sidel, Robin, and Chip Cummins. "Devon Energy to Acquire Mitchell in $3.1 Billion Cash-and-Stock Deal." *Wall Street Journal*, August 14, 2001. Accessed May 1, 2016. https://www.wsj.com/articles/SB997742296816559591.
Smith, Frank C. "It's Time Again to Stop the Bulldozers on the Bayou." *Houston Chronicle*, October 24, 2014. Accessed January 17, 2017. http://www.houstonchronicle.com/local/graymatters/article/Time-once-again-to-stop-the-bulldozers-on-the-5843294.php?cmpid=twitter-premium&t=22a454d7a879b87a02.
Solomon, Caleb, and Robert Johnson. "Lone Star Legend: One Tycoon in Texas Still Is Dreaming Big, Even If It's Out of Style." *Wall Street Journal*, July 28, 1993, A1. ProQuest (398328315).
Steward, Dan. "George P. Mitchell and the Barnett Shale." *Journal of Petroleum Technology*, November 2013, 60.
Tomkins-Walsh, Teresa. "Houston's Environmental Legacy: Terry Hershey, Community and Action." *Houston History* 10, no. 2 (Spring 2013), 4–5. Accessed January 15, 2017. https://houstonhistorymagazine.org/wp-content/uploads/2013/07/Hershey-environmental-legacy.pdf.
Wall Street Journal. "Mitchell Energy & Development Corp.: Advisers Are Hired to Look into Strategic Alternatives." October 7, 1999, B7. ProQuest (398789439).
———. "Mitchell Energy & Development Corp.: Firm to Stay Independent, Expects to Beat Forecasts." April 7, 2000, A6. ProQuest (398712425).
———. "Mitchell Energy & Development: Gas-Supply Pact with Unit of Occidental to Terminate." June 19, 1995, C15. ProQuest (03568660).
———. "Mitchell Energy Sells Properties." July 2, 1999, 1. ProQuest (39868330).
———. "Smelly Verdict—II." Unsigned editorial. June 12, 1996. https://www.wsj.com/articles/SB834533466553773500.
———. "Texas Court of Appeals Reverses Pollution Ruling." November 14, 1997, B6.
Williams, Lena. "Gladys Heldman, 81, a Leader in Promoting Womens Tennis." *New York Times*, June 25, 2003. Accessed December 5, 2016. http://www.nytimes.com/2003/06/25/sports/gladys-heldman-81-a-leader-in-promoting-womens-tennis.html.
Wilson, Robert R. "The Tevatron." *Physics Today*, October 1977, 23. Accessed March 5, 2017. http://dx.doi.org/10.1063/1.3037746.
Wise County Messenger. "Shuffling of Upstairs Personnel by C.M.&M. Co." May 28, 1959, 1. Newspapers.com.
Zimmerman, Ann. "A Lot of Gas." *Dallas Observer*, May 15, 1997. Newspapers.com.

Books

Boatman, Nicole T., Scott H. Belshaw, and Richard B. McCaslin. *Galveston's Maceo Family Empire: Bootlegging & the Balinese Ballroom*. Columbia, SC: History Press, 2014.

Bryce, Robert. *Power Hungry: The Myths of "Green" Energy and the Real Fuels of the Future*. New York: Public Affairs, 2010.
Carson, Rachel. *Silent Spring*. Boston: Mariner Books, 2002.
Cartwright, Gary. *Galveston: A History of the Island*. Fort Worth: Texas Christian University Press, 1998.
Dillon, David. *The Architecture of O'Neil Ford: Celebrating Place*. Austin: University of Texas Press, 1999.
Forsyth, Ann. *Reforming Suburbia: The Planned Communities of Irvine, Columbia, and The Woodlands*. Berkeley: University of California Press, 2005.
Galatas, Roger, and Jim Barlow. *The Woodlands: The Inside Story of Creating a Better Hometown*. Washington, DC: Urban Land Institute, 2004.
Gold, Russell. *The Boom: How Fracking Ignited the American Energy Revolution and Changed the World*. New York: Simon & Schuster, 2014.
Goodwin, Lawrence. *Texas Oil, American Dreams*. Austin: Texas State Historical Association, 1996.
The Greek Texans. San Antonio: University of Texas at San Antonio Institute of Texan Cultures, 1974.
Jackson, John A. *Wise County: 25 Years of Progress*. Dallas: Williamson Printing, 1972.
Kutchin, Joseph W. *How Mitchell Energy & Development Corp. Got Its Start and How It Grew*. Updated ed. Boca Raton, FL: Universal Publishers, 2001.
Larson, Erik. *Isaac's Storm: A Man, a Time, and the Deadliest Hurricane in History*. New York: Vintage Books, 1999.
Meadows, Donella, Jorgen Randers, and Dennis L. Meadows. *Limits to Growth: The 30-Year Update*. White River Junction, VT: Chelsea Green, 2004.
Mitchell, Christie. *The Night Owl*. With Bill Cherry. Galveston, TX: self-published, Bill Cherry and George P. Mitchell, 2011.
Mitchell, Johnny. *Energy—And How We Lost It*. Houston: Gulf Publishing, 1979.
______. *The Secret War of Captain Johnny Mitchell*. Houston: Gulf Publishing, 1976.
Morgan, George T., and John O. King. *The Woodlands: New Community Development, 1964–1983*. College Station: Texas A&M University Press, 1987.
Payne, Richard, and Geoffrey Leavenworth. *Historic Galveston*. Houston: Herring Press, 1985.
Payne, Richard, and Drexel Turner. *The Woodlands*. The Woodlands, TX: The Woodlands Development Corporation, 1994.
Pratt, Joseph. "George P. Mitchell, Chairman of Mitchell Energy & Development Corporation." In *The Oil Makers: Insiders Look at the Petroleum Industry*, edited by Jeffrey Share, 147–53. Houston: Rice University Press, 1995.
Presley, James. *A Saga of Wealth: The Rise of the Texas Oilmen*. Austin: Texas Monthly Press, 1983.
Rao, Vikram. *Shale Oil and Gas: The Promise and the Peril*. Research Triangle Park, NC: Research Triangle Institute Press, 2015.
Riordan, Michael, Lillian Hoddeson, and Adrienne W. Kolb. *Tunnel Visions*. Chicago: University of Chicago Press, 2015.
Schmandt, Jurgen. *George P. Mitchell and the Idea of Sustainability*. College Station: Texas A&M University Press, 2010.
Share, Jeffrey, ed. *The Oil Makers: Insiders Look at the Petroleum Industry*. College Station: Texas A&M University Press, 2000.

Sheehy, Sandy. *Texas Big Rich: Exploits, Eccentricities, and Fabulous Fortunes Won and Lost*. New York: William Morrow, 1990.
Simmons, Matthew. *Twilight in the Desert*. New York: John Wiley & Sons, 2005.
Steward, Dan. *The Barnett Shale Play: Phoenix of the Fort Worth Basin; A History*. Fort Worth: North Texas Geological Society, 2007.
Woods, Clinton, ed. *Ideas That Became Big Business*. Baltimore: Founders, 1959.
Yergin, Daniel. *The Prize: The Epic Quest for Oil, Money & Power*. New York: Simon & Schuster, 1991.
______. *The Quest: Energy, Security, and the Remaking of the Modern World*. New York: Penguin Group, 2011.
Zuckerman, Gregory. *The Frackers: The Outrageous Insider Story of the New Billionaire Wildcatters*. New York: Penguin Group, 2013.

Thesis

Miller, Reagan Scott. "The Architecture of MacKie and Kamrath." Master's thesis, Rice University, 1993.

Government Documents

Mitchell Energy and Development Corp. 10-K annual report for the fiscal year ended January 31, 1994. Filed with the US Securities and Exchange Commission, April 26, 1994. https://www.sec.gov/Archives/edgar/data/311995/0000950129–94–000319.txt.
Mitchell Energy and Development Corp. 10-K annual report for the fiscal year ended January 31, 1998. Filed with the US Securities and Exchange Commission, April 16, 1999. https://www.sec.gov/Archives/edgar/data/311995/0000950129–98–001454.txt.
Mitchell Energy and Development Corp. 10-K annual report for the year ended December 31, 2000. Filed with the US Securities and Exchange Commission, March 9, 2001. https://www.sec.gov/Archives/edgar/data/311995/000095012901001352/h84720e10-k405.txt.
"Strand Historic District, T. Jefferson League Building." Historic American Buildings Survey, Heritage Conservation and Recreation Service, US Department of the Interior. HABS No. TX-3296 F, 1–3. Accessed February 4, 2017.

Internet Sites

"Allen's Landing, TX." *Handbook of Texas Online*. Texas State Historical Association. https://www.tshaonline.org/handbook/online/articles/hvabg.
Barbee, Barbi. "The History of Buffalo Bayou in 1 Minute." Culture Trip, June 30, 2016. Accessed January 17, 2017. https://theculturetrip.com/north-america/usa/texas/articles/the-history-of-buffalo-bayou-in-1-minute/.
"Bush: Kyoto Treaty Would Have Hurt Economy." Associated Press via NBCnews.com, June 30, 2005. http://www.nbcnews.com/id/8422343/ns/politics/t/bush-kyoto-treaty-would-have-hurt-economy/#.W01aSbi1vRZ.
"Cushing, OK WTI Spot Price FOB." US Energy Information Administration. https://www.eia.gov/dnav/pet/hist/LeafHandler.ashx?n=pet&s=rwtc&f=m.
"Cynthia Woods Mitchell." Paid obituary, *Houston Chronicle*, December 30, 2009.

Accessed October 18, 2016. http://www.legacy.com/obituaries/houstonchronicle/obituary.aspx?pid=137965860.
"Dickson Gun Plant." *Handbook of Texas Online*. Texas State Historical Association. Accessed November 12, 2016. https://tshaonline.org/handbook/online/articles/dmd01.
"EOG Resources Elects Bill Stevens and Leighton Steward to Board of Directors." Company press release, July 2, 2004. Accessed May 9, 2016.
"George Mitchell Commits $25 Million to Giant Magellan Telescope." Giant Magellan Telescope, May 1, 2011. Accessed April 9, 2016.
"George P. Mitchell." Texas Tennis Museum and Hall of Fame. Accessed May 30, 2016. https://www.texastennismuseum.org/mitchell.
"George P. Mitchell '40," Texas A&M University Corps of Cadets Hall of Honor. Accessed February 15, 2016. http://corps.tamu.edu/portfolio-items/george-mitchell/.
"George P. Mitchell: Sixty Years of Tradition Continued." Texas A&M University press release, May 9, 2014. Accessed November 12, 2015. http://engineering.tamu.edu/news/2014/05/09/george-p-mitchell-sixty-years-of-tradition-continued.
"Henry Hub Natural Gas Spot Price." US Energy Information Administration. Accessed May 9, 2016. https://www.eia.gov/dnav/ng/hist/rngwhhdm.htm.
"Houston, TX." *Handbook of Texas Online*. Texas State Historical Association. Accessed December 7, 2016. https://www.tshaonline.org/handbook/online/articles/hdh03.
"Houston Metro Home Prices & Values," Zillow.com. Accessed March 30, 2017. http://www.zillow.com/houston-metro-tx_r394692/home-values/.
Hutchins, Shana. "Life Forces and Legacies." *Texas A&M Science Blog*, January 29, 2014. Accessed October 4, 2016. https://tamuscience.com/2014/01/29/life-forces-and-legacies/.
"Introducing Peloponnese." Lonely Planet. http://www.lonelyplanet.com/greece/the-peloponnese/introduction.
"John 'Jarrin' John' Kimbrough." National Football Foundation, College Football Hall of Fame. Accessed August 8, 2015. http://www.footballfoundation.org/Programs/CollegeFootballHallofFame/SearchDetail.aspx?id=30081.
Koppes, Steve. "Wendy Freedman, World-Leading Astronomer, Joins UChicago Faculty." *University of Chicago News*, August 7, 2014. Accessed April 5, 2016. https://news.uchicago.edu/article/2014/08/07/wendy-freedman-world-leading-astronomer-joins-uchicago-faculty.
"Logging Basics." *Crain's Petrophysical Handbook*. Accessed May 24, 2016. https://www.spec2000.net/01-whatisalog.htm.
MacRae, Michael. "Fracking: A Look Back." American Society of Mechanical Engineers. December 2012. Accessed January 6, 2016. https://www.asme.org/engineering-topics/articles/fossil-power/fracking-a-look-back.
"Mardi Gras History." Galveston.com. Accessed February 4, 2017. http://www.galveston.com/mardigrashistory/.
"Mitchell Energy and Development Corporation History." Funding Universe. Accessed January 4, 2017. http://www.fundinguniverse.com/company-histories/mitchell-energy-and development-corporation-history/.
"Mitchell Tennis Center." Texas A&M University, Quick Facts. Accessed December

5, 2016. http://www.12thman.com/facilities/?id=13.
"Natural Gas Expected to Surpass Coal in Mix of Fuel Used for US Power Generation in 2016." US Energy Information Administration. https://www.eia.gov/todayinenergy/detail.php?id=25392.
"Newton Bateman Memorial High School." Clio. Accessed January 4, 2017. https://theclio.com/web/entry?id=28258.
"Prestigious Stephen Hawking Chair in Physics Goes to Christopher Pope." Texas A&M University press release, October 28, 2002. Accessed February 29, 2016. http://www.science.tamu.edu/news/story.php?story_ID=13#.V_LQWJMrJ24.
"Robert Everett Smith." *Handbook of Texas Online*. Texas State Historical Association. Accessed January 3, 2016. https://tshaonline.org/handbook/online/articles/fsm57.
Rosen, Kamran. "Best Places to Live in Texas." Nerd Wallet, June 15, 2015. Accessed March 31, 2017. https://www.nerdwallet.com/blog/mortgages/best-places-to-live-in-texas-2015/.
Taylor, James. "Fracking, Lower Gasoline Prices Returned $1,000 to Household Budgets Last Year." Forbes.com, February 3, 2017. Accessed March 15, 2017. https://www.forbes.com/sites/jamestaylor/2017/02/03/fracking-lower-gasoline-prices-returned-1000-to-household-budgets-last-year/#73f8e66030ce.
"Underground Railroad." Jacksonville Area Convention & Visitors Bureau, 2014. Accessed January 4, 2017. http://jacksonvilleil.org/wp-content/uploads/2014/08/UGRR-Brochure2014.pdf.
"US Dry Natural Gas Production." US Energy Information Administration. https://www.eia.gov/dnav/ng/hist/n9070us2a.htm.
Warnock, John W. "Exploiting Saskatchewan's Potash: Who Benefits?" Canadian Centre for Policy Alternatives, January 2011. Accessed January 23, 2017. https://www.policyalternatives.ca/sites/default/files/uploads/publications/Saskatchewan%20office/2011/01/Exploiting%20SK%20Potash%2001–25–11.pdf.
Watts, Anthony. "USA Meets Kyoto Protocol Goal—Without Ever Embracing It." *Watts Up with That?* (blog), April 5, 2013. https://wattsupwiththat.com/2013/04/05/usa-meets-kyoto-protocol without-ever-embracing-it/.
"Watts Riots." Civil Rights Digital Library. Accessed January 24, 2017. http://crdl.usg.edu/events/watts_riots/?Welcome.
"What Is GMT?" Giant Magellan Telescope. Accessed April 7, 2016. http://www.gmto.org/overview/.
Yoquinto, Luke. "The Truth about Guar Gum." *LiveScience*, July 27, 2012. Accessed January 7, 2016. http://www.livescience.com/36580-guar-gum-weight-loss-cost.html.

Newspapers

Corsicana (TX) Daily Sun
Dallas Morning News
Galveston Daily News
Houston Chronicle
Houston Post
Oklahoman (Oklahoma City, OK)
Wall Street Journal
Wise County Messenger (Decatur, TX)

Index

The letter *p* following a page locator denotes a photograph.